Lecture Notes in Economics and Mathematical Systems

388

Dieter Bartmann   Martin J. Beckmann

Inventory Control

Models and Methods

Springer-Verlag

Berlin Heidelberg New York
London Paris Tokyo
Hong Kong Barcelona
Budapest

Authors

Prof. Dr. Dieter Bartmann
Institute of Information Management
University of St. Gallen for Business
Administration, Economics, Law and Social Sciences
Dufourstr. 50, CH-9000 St. Gallen

Prof. Dr. Martin J. Beckmann
Institute of Applied Mathematics and Statistics
Technical University of Munich
Arcisstr. 21, W-8000 München 2, FRG

ISBN 3-540-55820-9 Springer-Verlag Berlin Heidelberg New York
ISBN 0-387-55820-9 Springer-Verlag New York Berlin Heidelberg

Typesetting: Camera ready by author/editor
42/3140-543210 - Printed on acid-free paper

TP

FOREWORD

Inventory control is a major field in OR. Interest in the problems of optimal stock management at a scientific level goes back to the start of the 20th century. The most important impulse, however, came after the 2nd World War when scientists of the caliber of Jacob, Marschak, Kenneth Arrow, Samuel Karlin among others looked into the problem of optimal stocking under stochastic demand. It was characteristic of this discipline, that methods of solving problems of this type were developed first before the necessary commercial electronic data processing for their ready application were available.

The importance of inventory control in business increased dramatically with the increasing interest rates of the 70s. It was the rule of the hour to release surplus operating capital tied up in excessive inventories and to use the resulting liquidity to finance new investments. There arose a need for intelligent solutions to the problem. Unfortunately, OR experts and software applications developers were following separate paths. Hence, the opportunity to find the best solutions, using the combined potentials of theory, problem analysis and experience, was not exploited.

Today, we stand before the development and realization of sophisticated CIM concepts and it is imperative to reset the course. This book is a contribution to this effort. It shows how inventory control, with the help of OR, can be rationally structured. It is understandable that a complete treatment of this vast material is not possible and it is not intended. The book limits itself to standard models and the important ramifications. Special emphasis is given to fundamentals. On one hand, the reader is shown how the models are appropriately formulated and extended for special problems. On the other hand, the needed mathematical derivations are completely and comprehensively described so that, using the methods learned, the reader will be able to work out his own models not treated in this book.

Numerical methods for solving problems with random demand in complicated cases were given emphasis in this book. The important algorithms are thoroughly treated so that special situations can be handled.

The book is intended for OR practitioners and economists, as well as for information managers who are engaged in the development of modern computerized inventory systems, working for corporations, software houses or computer manufacturers.

The authors were actively supported in the writing of this book. Ingrid Riedlbeck and Susanne Spielvogel, both mathematics degree holders, patiently checked the mathematical derivations and reviewed them in detail. Mr. Robert Hackl read the proofs for the whole text and made the diagrams. Mrs. Karola Treiber and Mrs. Bernardy Schwarzwälder did the typing. Maria Luisa and Roberto Asuncion translated the German text, prepared the charts and typed the English manuscript. Finally, we would like to acknowledge the contribution of Prof. James Pope for his valuable comments and corrections to the English edition. To all we express our sincere gratitude.

St. Gallen/Munich, April 1992

Dieter Bartmann   Martin J. Beckmann

OVERVIEW

The book is divided into six chapters. Chapter 1 deals with inventory where demand is deterministic. Chapters 2–5 consider cases of stochastic demand. Chapter 6 is devoted to computational procedures.

CHAPTER 1 follows a historical line. The lot size model of Wilson (or Harris or Andler) is presented after a short introduction (Section 1). Although the assumptions in these models are of the simplest type, the derived formula for the optimal lot size proves to be rather robust in many practical situations, e.g., in the transition from a constant rate of demand to the Poisson demand (as will be shown in Chapter 2).

Costs and sensitivity are discussed in Section 3. It is shown that these are decreasing with scale when the ordering rule is optimal, i.e., with increasing turnover, the cost per unit of inventory becomes smaller. Under decentralization, this effect of increasing returns to scale is partially lost. The appropriate formulas are derived. Using sensitivity analysis, the effects of various parameters on the total expected cost are discussed; first, when the rate of demand or the various costs are wrongly estimated; secondly, when an optimal order quantity is not realizable due to special packaging units or container size conditions; or, thirdly, when the desired period length between orders is predetermined because of internal business organizational reasons or because of prescribed delivery dates as in the case of the pharmaceutical industry.

The next two sections, Section 4 and Section 5, deal with multi–item inventories. Section 4 discusses the theoretical foundation of a classification for an "ABC–analysis" with respect to turnover volume and prices. Section 5 considers the question of stock maintenance. How high must the demand rate be in order to make it worthwhile to keep an article in stock at all?

To apply scientific inventory control it is important to have, as much as possible, an estimate of demand. Unfortunately, sales figures are often aggregated (monthly, quarterly, yearly) and disaggregated values are sometimes not readily available. It is, therefore, shown in Section 6 how the demand rate can be estimated from order data.

How does the optimal inventory policy change when the objective of the firm is profit maximation rather than minimization of inventory costs? In this connection, the question also comes up of how the inventories of a firm should be evaluated. These questions are examined in Sections 7 and 8.

The standard model needs modification in cases of quantity discounts. Two cases are discussed in Section 9: (a) the quantity discount is given only to the extent that the order exceeds a cutoff point and (b) it is granted for the whole order quantity when the cutoff point is exceeded.

In Section 10, we examine when a collective order is more advantageous than single orders. Up to this point, sales inventory or raw materials inventory have been discussed. In Section 11 production inventory or finished goods inventory for internal production is considered . How big is the optimal lot size in production with a continuous demand at a constant rate?

The consequences of inventory shortages are discussed in Section 12. For businesses with a monopolistic character, demand is not lost even when there are delivery bottlenecks (the so—called backorder case). Even then, however, an inventory deficit will cost something since profits can only be realized later. It turns out that shortages may be perfectly profitable. The optimal ordering cycles and order quantities can then be calculated.

Discrete lot sizes are handled in Section 13. This is especially important for small lots and for goods with low demand.

In Section 14, warehouse shelf space is considered. In the first case, a fixed shelf space is reserved for each good. In the second, the order period between two goods is staggered in such a way that the maximum shelf space is kept as small as possible. In addition to space limitations, budget constraints can be in effect. The question of limited space and/or budget limitations on the optimal order quantity is considered in Section 15.

In Section 16, we consider a "rolling demand horizon" where only the demand in the next period is exactly known but unknown thereafter. In Section 17, a fixed delivery schedule is considered. Once more the question arises as to when it is worthwhile to keep goods in stock and when it is more advantageous to sell on order.

In Section 18, the random deviations from a fixed delivery schedule are considered and an optimal system of "Just–in–Time" is determined.

CHAPTER 2 extends the simple Wilson inventory model to cases of random demand. A first example is a Poisson process. (Further generalizations, e.g., to random delivery schedules or distributed demand, are treated in Chapter 4). Sections 19 and 20 give an introduction to the Poisson process together with some generalizations and present the decision criterion to be used for decision making under risk. Section 21 deals with continuous interest payment and infinite payment flows. In Sections 22 and 23, the inventory model with Poisson demand in the discounted and non–discounted case is formulated as a Dynamic Program using Bellman's "Principle of Optimality". The model is further generalized in Section 24 to handle the case of randomly dependent demand, a Semi–Markov process. Section 25 demonstrates the use of policy iteration of dynamic programming to show that even for stochastic demand, the optimal order quantity is identical to the Wilson lot size of the deterministic model.

In CHAPTER 3, single period models are discussed. This type of "inventory" problem occurs, for example, in fashion articles or ticket sales or in preparing for an expedition. The basic model known as the newsboy problem is presented in Section 26. In this connection, we determine when it is profitable to enter into a single period business at all.

The dependence of the optimal lot size on the parameters of the demand distribution and on the inventory and shortage costs are discussed in Section 27. With the use of entropy, it is shown that the single period cost increases as the difference between inventory and shortage costs becomes smaller. In Sections 28 and 29, we look into the optimal period length. The "overbooking of reservations" is similar to the newsboy problem. Since it seldom happens that all reservations are actually claimed, it pays off to sell a certain part of a reserved quota a second time.

CHAPTER 4 treats the stochastic demand model under continuous monitoring. Two approaches are dynamic programming and the Method of State Probability. The latter is explained in Section 31 and is applied to a model with geometrically distributed demand and to a model with Poisson demand and exponential delivery time. Here we also consider the possibility that the inventory cost depends on the maximum stock level. This occurs, for instance, when storage space must be rented.

The emphasis of Sections 33–35 is on delivery times. Supplier reliability is an important factor under competition. In many cases one can therefore look at the delivery time as certain. In Section 33, a model with fixed delivery time is discussed. In a monopolistic situation or where the goods are allocated, the uncertainty occurs not so much in demand but in the delivery times. This is especially observable in developing countries. The situation of internal production or just–in–time deliveries is specially treated here because delays in delivery are critical.

CHAPTER 5 examines stochastic inventory models with periodic monitoring. Even though continuous inventory monitoring is now a common practice, many businesses still use periodic inspection and order decisions.

Periodic models occur also when arrangements have been made with suppliers for deliveries at specific times. The basic Arrow–Harris–Marschak (AHM) model with a finite (Section 36) and infinite (Section 37) planning horizon is formulated. This model is normalized to a standard format in Section 38. Optimal order policies for different expected values and standard deviations of the demand distribution can then be derived directly from the optimal solution of the standard model.

A special model (Section 39) is the AHM–Model with exponentially distributed demand, the period analog to the continuous model with Poisson demand. It is solved explicitly.

The optimality of the (s, S)–policy is examined in Sections 40–44. A method of calculating s and S is given for a special case. In Section 45 the model with a single period delivery lag is formulated. It is shown how this model fits into the framework of the AHM–Model. An interesting result is that the inventory fluctuation of a model with positive delivery time is greater than for models with instantaneous delivery.

In general, a lengthening of the delivery time makes inventory control more expensive.

The normal assumption of a stationary demand process is not always realized in practice; demand levels are subject to fluctuations over time. Information about the immediate past may be available from which a short term forecast can be made. This information should be considered in the model. This is done in the following two sections. Autocorrelated demand is assumed in Section 46. In Section 47, endogeneous and exogeneous forecast mechanisms are introduced into the model. Exponential smoothing is given as an example.

Special considerations are given to goods which have a normally distributed demand, have small rates of market growth and whose turnover is forecast with the help of exogeneous variables, such that successive forecasts are not autocorrelated. If, for example, the exogeneous variable is given by the change in the gross national product, then this approach is suitable for goods which are subject to the acceleration principle, e.g. investment goods or spare parts.

So far we have attempted to derive formulas for the optimal lot size and ordering rules. In cases where this is not possible, one can lean on the computational method of dynamic programming. This is the content of Chapter 6. In Section 48, the method of value iteration is developed. It is the most general method of dynamic programming and can be applied to inventory models which, due to their complicated cost structure, deviate too much from the studied basic models. The advantages and weaknesses of this method are shown and a possibility of shortening the computation time is given.

Policy iteration is presented in Section 49. It is an alternative to value iteration in the inventory problem with infinite planning horizon. For this type of problem, value and policy iteration can be combined into a third method, the so-called policy-value iteration. This is, however, not considered here because the method of bisection in connection with dynamic programming introduced in Section 50 promises to be better.

In the last section, the AHM–Model is specially considered for the backorder case without discounting. For this model, a standard form was derived (Section 38) which is, however, subject to a limitation with respect to the assumed distribution of demand. It is, therefore, important that fast methods of computation are available for models with a general demand distribution. One such method was developed by Federgruen and Zipkin. It is discussed in Section 51.

TABLE OF CONTENTS

CHAPTER 3:   STOCHASTIC SINGLE PERIOD MODELS

CHAPTER 4:   STOCHASTIC MODELS WITH CONTINUOUS REVIEW

CHAPTER 5:   STOCHASTIC MODELS WITH PERIODIC REVIEW

CHAPTER 6:   NUMERICAL METHODS

CHAPTER 1:
DETERMINISTIC INVENTORY MODELS

§1   INTRODUCTION

J.M. KEYNES differentiated three motives for holding money which can be applied to inventory problems.

1.   The Transaction Motive
      Since outflows are not synchronized with inflows, stocks are needed to bridge these discrepancies. Usually, incoming goods arrive in greater quantities and in longer time intervals than outgoing ones.

2.   The Precautionary Motive
      If an order is placed, one must maintain reserve stocks in order to satisfy the demand while awaiting delivery.

3.   The Speculative Motive
      If prices are expected to rise, it pays to keep stocks on hand.

In operations research (OR), inventory is typically geared towards the first two motives. The third is occasionally treated in linear optimization as the so–called warehousing problem.

Inventory theory belongs to the first and, therefore, classical application areas of OR. It was strongly supported in the 1950s, primarily by the US Navy.

Scientists of the caliber of OSKAR MORGENSTERN, JAKOB MARSCHAK, KENNETH ARROW, HERBERT SCARF, THOMAS WHITIN, JACK KIEFER and others have intensively worked with the application of OR and Statistics on inventory problems. Its beginnings, however, go back to the mythical WILSON much earlier at about the turn of the century. At that time, the question of optimal inventory control strategies was a long disputed issue. This led to the development of the Theory of Dynamic Programming (RICHARD BELLMAN).

§2   ECONOMIC ORDER QUANTITY (EOQ)

The standard case of the EOQ (or optimal lot size) problem is a trading firm which orders a good to store up stocks. Customer demand is satisfied by stocks on hand. We assume a constant rate of demand. Let

λ :   Rate of Demand
y :   Stock on Hand

The assumption of a constant rate of demand is a highly ideal one in wholesale and retail transactions. On the other hand, it occurs very frequently in raw material storage in continuous or batch type production.

Cost Structure of the Inventory Model

Ordering Cost: We assume a linear relationship for the ordering cost (Fig. 2.1).

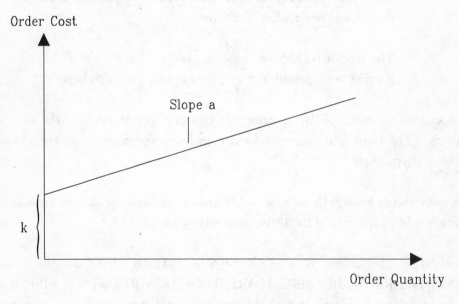

Figure 2.1: Ordering cost curve

k :   fixed ordering cost. This covers administrative costs (e.g., $10–50; a business letter costs approximately $7), customer complaints, etc.

a :   proportional ordering cost, e.g., transportation cost, cost for controlling incoming goods; in our model, it is mainly the buying price.

<u>Inventory Cost</u>: It consists of interest costs, handling costs and rental costs for storage (even if one is the owner of the warehouse; in this case, the rental cost is an opportunity cost; the possibility of using the warehouse for other purposes is given up). Moreover, cost of wastage (in India, 1/4 of the grain harvest is eaten up by rats), depreciation and obsolescence can also arise. All these costs are summed up in the inventory cost.

h :   Inventory cost per unit item and unit time (Inventory cost rate)

<u>Shortage Cost</u>: Shortages occur in case stocks are very low and, therefore, demand cannot be fully satisfied. These are charged penalty costs.

g :   Shortage cost per unit item and unit time
z :   Amount of shortage (deficit)
G :   Shortage cost

Usually, shortage costs are assumed to be proportional to the amount of shortage,

$$G = g \cdot z.$$

Shortage costs can also be thought of as being independent of the size of the deficit z,

$$G = g \cdot \delta(z), \qquad \delta(z) = \begin{cases} 0, \text{ for } z = 0, \\ 1, \text{ for } z > 0. \end{cases}$$

δ is the so–called Kronecker delta. This second manner of calculating shortage costs was used, for example, by the US Navy. The inventory problem consisted of determining how many (replacement) parts a ship should carry to cover its needs during a voyage. Replenishment at sea happens infrequently. If the required number of spare parts of the same type is more than the number on board , then it is insignificant how many parts are lacking. High costs are incurred even if only a single part is missing.

Shortages can arise if the stock is not permanently recorded (periodic inspection), if the stock is ordered too late or when the ordered quantity is delivered late.

The cost structure described here is very simple. A detailed discussion of differentiated cost considerations can be found, however, in business economics literature.

The WILSON Lot Size Formula (also ANDLER's Formula or HARRIS' formula)

We consider the simple case of inventory with the above cost structure, constant demand rate and permanent stock control. Shortages are not allowed (see §10). The stock level is controlled by the following operational characteristic (Fig. 2.2)

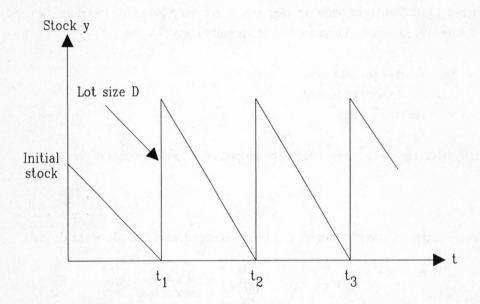

Figure 2.2: Operational characteristic of inventory control

Since the delivery time is zero, it is obvious that an order only pays off when inventory reaches zero ($t = t_1$). The order quantity is

D: Lot Size.

When the inventory level is again zero $(t = t_2)$, a second order is placed. Because $\lambda =$ constant, the system is stationary and there is no reason to choose another order quantity different from the first one. Since the situation at period t_1 is the same as that in time t_2, the optimal quantity in t_1 must, therefore, also be optimal in t_2.

The demand rate λ is pre–determined, i.e., independent of our actions. Hence, the area of optimization lies in the lot size. We need to determine the minimum cost order quantity.

The objective function "Cost per cycle $(t_i - t_{i-1})$" is not suitable since minimizing these costs

$$k + aD + h\, \frac{D}{2} \cdot \frac{D}{\lambda} \underset{D}{\to} \text{Min}$$

average inventory $\underline{\qquad}$ $\underline{\qquad}$ cycle length $t_i - t_{i-1}$

leads to the absurd result: optimal lot size $D^* = 0$.

A possible objective function is the average unit cost

$\overline{C}$:   Average unit cost

$$\overline{C} = \frac{k + aD + h \cdot \frac{D}{2} \cdot \frac{D}{\lambda}}{D} \underset{D}{\to} \text{Min} \; .$$

Another possible objective function is the cycle cost per unit time

C: Cost during a cycle per unit time.

$$C = \frac{k + aD + h \cdot \frac{D}{2} \cdot \frac{D}{\lambda}}{\frac{D}{\lambda}} \underset{D}{\to} \text{Min} \; . \tag{2.1}$$

Because of the proportionality $C = \lambda \bar{C}$ and λ = constant, it is irrelevant whether we use $\bar{C}$ or C. Both are convex in D.

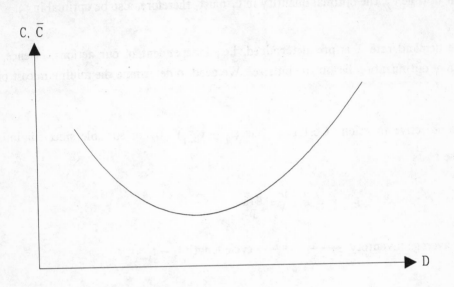

Figure 2.3: Convex objective function C or $\bar{C}$

Therefore, we obtain the optimal lot size D^* by differentiating the objective function C

$$\min_{D} C(D) \iff \frac{dC}{dD} = 0$$

$$\frac{dC}{dD} = 0: \quad -\frac{k\lambda}{D^2} + \frac{h}{2} = 0$$

$$\Rightarrow \quad \boxed{D^* = \sqrt{\frac{2\lambda k}{h}}} \quad . \tag{2.2}$$

Equation (2.2) is called the WILSON Lot Size Formula or the HARRIS Formula.

This is a practical equation as shown by a short sensitivity analysis. The optimal lot size D^* increases with an increasing demand rate λ as well as with an increasing fixed ordering cost.

Interval between Two Orders
Let

T: Interval between two orders

From (2.2) one immediately derives

$$T = \sqrt{\frac{2k}{\lambda h}} \; . \tag{2.3}$$

Average Range of Coverage
An important reference value is the inventory–sales ratio. It tells something about the long–term efficiency of a stock control system. For an optimal ordering policy

$$\frac{\text{Average Inventory}}{\text{Average Sales}} = \frac{D^*}{2\lambda} = \sqrt{\frac{k}{2h\lambda}} \; . \tag{2.4}$$

Research has shown that, in spite of operations research, the average inventory holdings in the last two decades have increased. There are two reasons for this:

1.   Wages and salaries have increased sharply so that the rate k/h increased in spite of increasing interest costs (h increases) and the decrease in part of the fixed costs through EDP.
2.   Decentralization has increased the number of warehouses. Moreover, type variants (product variations) have increased such that the demand rate per variant and stock location have decreased which according to (2.4) leads to an increase in the average range of coverage.

Returns to scale can also be seen from (2.4). The inventory–sales ratio of a company becomes more favorable with increasing sales volume. This, however, says something about the cost. The following section deals with cost considerations.

§3   COSTS AND SENSITIVITY

Costs

The cost function C from equation (2.1) has among others a proportional ordering cost $\lambda \cdot a$. It can be seen that this term does not influence the optimal lot size. Over a long period of time, these ordering costs cannot be avoided and they can be considered to be fixed costs. Therefore these non–influential terms are ignored in optimization. Let the new cost function without the ordering cost be

$$c = C - \lambda a .$$

For a period of length t

$$c = \frac{k}{t} + \frac{hD}{2} . \tag{3.1}$$

For an optimal ordering quantity D^*

$$c = \sqrt{2k\lambda h} . \tag{3.2}$$

As is to be expected, the cost of an order cycle per unit time increases with increasing sales volume. The increase, however, is sublinear:

$$c \sim \sqrt{\lambda} .$$

For the unit cost $\bar{c} = c/\lambda$ per time unit

$$\bar{c} = \sqrt{\frac{2kh}{\lambda}} . \tag{3.3}$$

It decreases, therefore, with increasing turnover.

$$\bar{c} \sim \frac{1}{\sqrt{\lambda}} .$$

This also shows the effect of increasing returns to scale (an advantage of bigger firms). The reason for this is INDIVISIBILITY, in our case, the indivisibility of the fixed order costs. Whether the order quantity is small or large, the fixed order costs, remain the same.

In many large firms, however, inventory is decentralized. Because of this, the scaling effect is partly lost as shown in the following consideration. With m warehouses a single warehouse will have a demand rate of λ/m. Let the total demand be λ. With decentralization, the total cost per cycle is then

$$m \sqrt{2kh\lambda/m} = \sqrt{m} \, c \, ,$$

i.e., larger than centralization by a factor of $\sqrt{m}$. Decentralization has often been justified in the course of corporate history and, hence an energetic drive is needed to break out of the imposing structures and to reorganize the logistics. A similar stimulus in the form of high interest rates happened at the beginning of the eighties as one tried hard to rationalize the sales income of a corporation to free liquid resources. This resulted in the centralization of inventory in many firms.

One should not ignore, however, the advantage of decentralization: improved customer service. This is not expressed in the above formulas (3.2), (3.3). For instance, they do not include the cost of transportation.

Sensitivity
The partial derivative $\frac{\partial c}{\partial x}$ gives information about how a change in each variable x affects the cost c. It is

$$\frac{\partial c}{\partial \lambda} = \sqrt{\frac{kh}{2\lambda}} \; ; \quad \frac{\partial c}{\partial k} = \sqrt{\frac{\lambda k}{2k}} \; ; \quad \frac{\partial c}{\partial h} = \sqrt{\frac{\lambda k}{2h}} \, .$$

These values, however, are dependent on the chosen units. One of the more important ratios is the elasticity ϵ. It measures the relationship of the relative changes of two quantities

$$\epsilon_{c,\lambda} = \frac{\frac{\partial c}{c}}{\frac{\partial \lambda}{\lambda}} \cdot \tag{3.4}$$

The elasticity can also be represented as a logarithmic derivative

$$\epsilon_{c,\lambda} = \frac{\partial \ln c}{\partial \ln \lambda} \cdot$$

The elasticity of c in relation to k and h are

$$\epsilon_{c,k} = \frac{\partial \ln c}{\partial \ln k} \quad \text{and} \quad \epsilon_{c,h} = \frac{\partial \ln c}{\partial \ln h},$$

respectively.

With $c = \sqrt{2k\lambda h}$, one obtains

$$\boxed{\epsilon_{c,\lambda} = \epsilon_{c,k} = \epsilon_{c,h} = \frac{1}{2}} \quad .$$

The elasticity of cost per period in relation to λ, k, h is always $\frac{1}{2}$. If, for example, the cost of k or h increases by p%, then the total cost c per period increases by $\frac{p}{2}$%. The same is true for the cost unit $\bar{c}$. It is important to be sure about the sensitivity of c and $\bar{c}$, because it can rarely be assumed in practice that k and h can be exactly determined.

The cost sensitivity in relation to changes in lot sizes, $\frac{\partial c}{\partial D}$, is also interesting. It is not always possible to realize the minimum cost c. Some of the reassons are technical conditions (container size, truck or tank capacity), special packaging or an arbitrary order cycle: week, month or quarter. Let us denote for a moment the optimal value with an asterix (*). With the help of the Taylor expansion about c*, we compute the cost difference c − c*. It is

$$c = c(D) = \frac{k\lambda}{D} + \frac{hD}{2} \qquad\qquad\qquad \text{(compare (3.1))}$$

$$\frac{\partial c}{\partial D} = -\frac{k\lambda}{D^2} + \frac{h}{2}$$

$$\frac{\partial^2 c}{\partial D^2} = \frac{2k\lambda}{D^3} \; ,$$

and, hence,

$$c - c^* = 0 + \frac{(D - D^*)^2}{2} \cdot \frac{2k\lambda}{(D^*)^3} + \dots$$

$\quad\quad\quad\quad \upharpoonleft$ the linear term disappears, since $\left.\dfrac{\partial c}{\partial D}\right|_{D^*} \overset{!}{=} 0$

Ignoring higher terms we have

$$c - c^* = \lambda k \frac{(D - D^*)^2}{(D^*)^3} \; .$$

How much it amounts to must be checked in each case.

Example:
Let k = $8, h = $0.01/day and unit, λ = 1 unit/day. Then $D^* = \sqrt{2\lambda k / h}$ = 40 units. This lot size is enough for 40 days. The cost c* per day is $c^* = \sqrt{2k\lambda h}$ = $0.40.

The good, however, can only be obtained in minimum lot sizes of 50 units: D = 50. How much is the increase in cost per day?

$$c - c^* \approx \lambda k \frac{(D - D^*)^2}{(D^*)^3} = \$\frac{1}{80} = \$0.0125.$$

This is an overestimate. The actual cost difference where c is computed according to (3.1) amounts to \$0.01. This means that $\Delta c/c = 2.5\%$ for every change $\Delta D/D$ of 20%. Average deviation from the optimal lot size is therefore not very noticeable. Reason: Δc is in the first approximation quadratic in ΔD.

§4   RM–SYSTEMS (ABC ANALYSIS)

The abbreviation RM stands for the latin "reductio ad maximum". In an RM–system, the goods are arranged according to importance. The importance of good i is considered according to its sales volume $\lambda_i a_i$, measured in terms of its buying price (and not by its selling price for which we measure costs). In our model, the buying price is the proportional ordering cost a_i.

Earlier research has shown that the goods may be roughly classified into three classes

Class	Number of Items	Approximate $\sum_i \lambda_i a_i$
A	20%	65%
B	40%	27%
C	40%	9%

This type of classification is popularly known as ABC–Analysis.

Is λa actually the right criterion for a classification based on cost? The cost function is $c = \sqrt{2k\lambda h}$. Accordingly, the criterion should be λkh. If, however, k is constant for all goods and $h \sim a$ (interest cost!) then

$$\lambda a \sim \lambda kh.$$

This gives the theoretical justification to use λa as a measure for the cost which causes one to hold inventory.

The purpose of an ABC classification is to save inventory management costs. Only goods of class A (highest importance) are handled according to the best possible methods. Note: In this case, continuous stock control is often necessary! The simplest inventory models are used for goods in classes B and C . One normally relies on the rule of thumb.

§5   PRODUCT–MIX DECISION

An ABC–Analysis can lead to the decision to straighten out the product mix and to discard certain articles, such as goods with high costs or low demand, the so–called slow–movers.

Let

p_i :   Selling price per unit of good i

a_i :   Buying price per unit of good i.

Then the profit per order period of length T is

$$\lambda_i T(p_i - a_i) - k_i - h_i D_i \frac{T}{2} .$$

The Rate of Profit G_i = Revenue minus Cost per period

$$G_i = \lambda_i(p_i - a_i) - \frac{k_i + h_i D_i \frac{T}{2}}{T}$$

$$= \lambda_i(p_i - a_i) - \sqrt{2k_i\lambda_i h_i} .$$

The optimal product mix is given by all goods with a positive rate of profit. The cutoff $\overline{\lambda}_i$ of demand, when $G_i = 0$, is called the BREAK–EVEN POINT. For $\lambda_i < \overline{\lambda}_i$, good i lies in the loss area, for $\lambda_i > \overline{\lambda}_i$ in the profit area.

The break–even point is set at $\overline{\lambda}_i = \dfrac{2k_i h_i}{(p_i - a_i)^2}$.

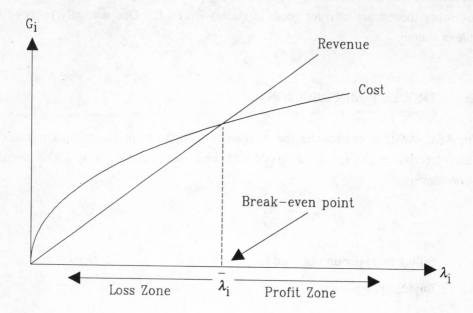

Figure 5.1: Break–even analysis

A systematic product mix adjustment is often done, for example, in the case of booksellers. If the rate of sales falls below a critical value, the books will not be republished and the remaining stocks will be sold cheaply. To avoid the danger of selling out too early, it is important to know λ_i exactly.

§6   ESTIMATING THE RATE OF DEMAND

Sales data in disaggregated form (not monthly, quarterly or yearly sales figures) are not always available. Therefore, we use previous order data to estimate the rate of demand. Let

t_i :   Interval between the last $(i+1)^{th}$ and i^{th} order

(Note: it is counted from the past, i.e., t_i is the ith period in the past)

D_i :   Order quantity at the last i^{th} order

      (Replenishment order!)

l_i :   Last i^{th} auxiliary value for λ; $l_i = D_i/t_i$

      (Note: The order D_i is the substitute for the demand before the last ith order)

In the lot size model it is assumed that λ is constant. Hence, it has to be tested whether the observations actually support the assumption. A quick answer can be obtained from a visual inspection of the series $\{l_i\}_{i \, \epsilon \, \mathbb{N}}$.

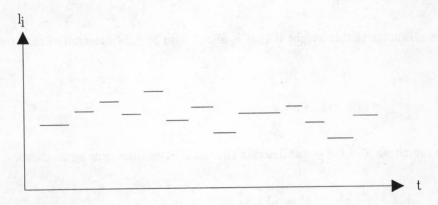

Figure 6.1: Time series of observations l_i

If the observations l_i as is shown in Fig. 6.1 fluctuates around a long–run constant average, then the arithmetic mean from the n existing observations is a suitable estimator for the true λ

$$\hat{\lambda} = \frac{1}{n} \sum_{i=1}^{n} l_i . \tag{6.1}$$

If one chooses only each of the last m observations, m is fixed, and one speaks of a moving average.

If the observations stretch over a longer period, shifts in the demand process, as a rule, will occur because of product mix changes, changes in brand loyalties, etc. It is then meaningful to give more weight to newer data than to old ones. By geometric weighting, one obtains for $n \to \infty$:

$$\hat{\lambda} = \frac{\sum\limits_{i=1}^{\infty} \rho^{i-1} l_i}{\sum\limits_{i=1}^{\infty} \rho^{i-1}} = (1-\rho) \cdot \sum\limits_{i=1}^{\infty} \rho^{i-1} l_i, \quad |\rho| < 1. \tag{6.2}$$

ρ is the weight factor.

The advantage of this weight is that it allows $\hat{\lambda}$ to be easily computed recursively. It is

$$\hat{\lambda}_{t+1} = (1-\rho) l_1 + \rho \hat{\lambda}_t, \qquad t = 1,2,\ldots \tag{6.3}$$

We substitute ρ with $1 - \rho$ and obtain the usual terms from time series theory

$$\boxed{\hat{\lambda} = \rho l_1 + (1 - \rho) \hat{\lambda}_1} \ . \tag{6.4}$$

l_1 is the last observation, $\hat{\lambda}_1$ the previous and $\hat{\lambda}$ the new estimate for λ. The equivalence between (6.2) and (6.3) is easily shown by successive solution of the recursion (6.3).

The estimation procedure (6.3) or (6.4) is called a first–order exponential smoothing. The previous values are exponentially damped. Through this, the speed of adaptation to a sudden occurrence of a change in level is increased considerably as compared to the arithmetic mean method. It becomes specially clear if one formulates equation (6.1) recursively

$$\hat{\lambda}_{t+1} = \frac{1}{t+1} l_1 + \frac{t}{t+1} \hat{\lambda}_t \tag{6.5}$$

and let t be very large. The latest observation is taken as the new estimate value with a weight of $1/(t+1)$. As time increases, this influence always becomes smaller. On the other hand, it remains constant with exponential smoothing.

The theoretical basis of the first order exponential smoothing lies in the modelling of an adaptive expected behavior according to the formula

$$E\{\lambda_{t+1}\} - E\{\lambda_t\} = \rho(l_1 - E\{\lambda_t\}) \quad ,$$

from which

$$E\{\lambda_{t+1}\} = \rho l_1 + (1-\rho)E\{\lambda_t\} \tag{6.6}$$

It describes the structure of the time series which fluctuates about a constant level, where this level is itself disrupted by random displacements.

$E\{\cdot\}$ is the Expected Value Operator

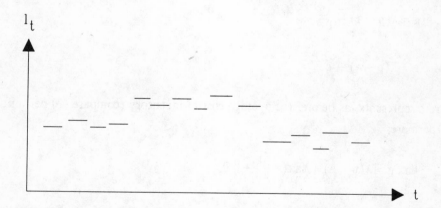

Figure 6.2: Time series with level displacements

Exponential smoothing is an appropriate forecasting method with this type of time series. Time series theory, moreover, gives a general statement about the structures of time series for which this forecasting method is optimal. More discussions on this and other refined variations of exponential smoothing can be found in BOX/JENKINS (1976) and MAKRIDAKIS/WHEELWRIGHT (1987).

The usual values of ρ lie between 0.01 and 0.1. The choice of a suitable value ρ is again in itself a decision problem where the perception about the speed of adaptation comes into play.

§7   PROFIT MAXIMIZATION

We assume that a good is sold at a price p per unit, bought at price a and the other data as before. The objective is profit maximization. The average profit per unit time amounts obviously to

$$g = \frac{pD - aD - k - h\,\frac{D}{2} \cdot \frac{D}{\lambda}}{D/\lambda} \,. \tag{7.1}$$

If the revenues and costs occurring within an inventory period are divided by the length of the period, then

$$g = \lambda(p - a) - \frac{\lambda k}{D} - \frac{h}{2}D$$

$$g = \lambda(p - a) - c \tag{7.2}$$

where c represents, as before, the average cost of inventory (compare §3) per unit time. Furthermore,

$$\underset{D}{\text{Max}}\ g = \lambda(p - a) + \underset{D}{\text{Max}}\left(-\frac{\lambda k}{D} - \frac{h}{2}D\right)$$

$$= \lambda(p - a) - \underset{D}{\text{Min}}\left(\frac{\lambda k}{D} + \frac{h}{2}D\right) \,. \tag{7.3}$$

The profit maximization problem is therefore identical to the cost minimization problem of the standard inventory theory except for the additive constant $\lambda(p-a)$.

§8   INVENTORY EVALUATION

A firm has a license to engage in the warehousing business until time period T. Let the current stock be y at a given period t. How large is the commercial value of the firm? In other words, how does one evaluate the inventory y?

The value of the firm is obviously a function of the stock level y as well as the remaining time T–t. It is described by

$$v(y, T - t)$$

During a short period Δt, it evolves as follows

$$v(y, T - t) = p \lambda \Delta t - hy\Delta t + v(y - \lambda \Delta t, T - t - \Delta t) , \quad y > 0 \qquad (8.1)$$

since the current revenue is $p\lambda \Delta t$, the current costs are $hy\Delta t$ and stocks are reduced by $-\lambda \Delta t$.

If y = 0, then

$$v(0, T - t) = -k - aD + v(D, T - t) , \quad y = 0 \qquad (8.2)$$

applies because stocks must be replenished up to D and that causes the cost k + aD.

The Taylor–Approximation for $v(y - \lambda \Delta t, T-t - \Delta t)$ is

$$v(y - \lambda \Delta t, T - t - \Delta t) = v(y, T - t) - v_y \cdot \lambda \Delta t - v_t \cdot \Delta t$$

Substituting in (8.1) and dividing by Δt result in the partial differential equation for v

$$\lambda v_y + v_t = \lambda p - hy \qquad (8.3)$$

with the boundary condition (8.2) and end value condition

$$v(y, 0) = 0 . \tag{8.4}$$

So that the end condition will hold, assume that

$$y(T) = 0$$

i.e., a final stock of zero is planned.

It is not unreasonable to attempt to separate the value function into a purely time–dependent and a purely volume–dependent part

$$v(y, T - t) = w(y) + g \cdot (T - t) . \tag{8.5}$$

In addition, the time–dependent part is set proportional to the remaining time. The proportionality factor is to be interpreted as the rate of profit per unit time. Using the formula (8.5), the partial differential equation (8.3) yields

$$\lambda w'(y) + g = \lambda p - hy. \tag{8.6}$$

Integrating from 0 to y gives

$$w(y) - w(0) = (p - \frac{g}{\lambda}) y - \frac{h}{2\lambda} y^2 . \tag{8.7}$$

In particular,

$$w(0) = 0 .$$

Using the boundary condition (8.2), one obtains for y = D,

$$w(D) - w(0) = k + aD = (p - \frac{g}{\lambda}) D - \frac{h}{2\lambda} D^2 .$$

The rate of profit is then determined to be

$$g = \lambda [p - a - \frac{k}{D} - \frac{h}{2\lambda} D] . \tag{8.8}$$

The profit margin with rate λ is

$$p - a - \frac{k}{D} - \frac{h}{2\lambda} D = p - a - \bar{c}$$

$\bar{c}$ is the unit cost per time (compare (3.3)). Substituting (8.8) and $w(0) = 0$ in (8.7) results into the value of inventory y

$$w(y) = [a + \frac{k}{D} + \frac{h}{2\lambda} D] \, y - \frac{h}{2\lambda} y^2 \; . \tag{8.9}$$

The value of the business consists of the value of inventory (8.9) and the value of the remaining time g (T–t). The value of the stock is a quadratic, not a linear nor a proportional, function of the stock. It reaches its maximum at

$$y^* = \frac{\lambda(a + \frac{k}{D} + \frac{h}{2\lambda} D)}{h} \; .$$

$$y^* = \lambda \frac{a}{h} + D \tag{8.10}$$

using the Wilson lot size formula for D. The value of the stock increases therefore with the stock in the whole range $0 \le y \le D$.

If one considers only the added value of stock m(y), i.e., the surplus above the buying price a, then according to (8.9) we have

$$m(y) = (\frac{k}{D} + \frac{h}{2\lambda} D) \, y - \frac{h}{2\lambda} y^2$$

$$m(y) = \sqrt{\frac{2kh}{\lambda}} \cdot y - \frac{h}{2\lambda} y^2 \; . \tag{8.11}$$

This added value assumes its maximum if

$$\frac{dm}{dy} \overset{!}{=} 0 \; , \quad \Rightarrow \quad \sqrt{\frac{2kh}{\lambda}} - \frac{h}{\lambda} y = 0$$

$$\Rightarrow y = \sqrt{\frac{2k\lambda}{h}} = D \; .$$

The optimal order quantity is therefore the one which maximizes the added value of inventory.

The evaluation of inventory levels and its clear delineation from the time value of a business enterprise are relevant economic problems (GRUBBSTRÖM).

§9   QUANTITY DISCOUNT

A modification of the standard model is necessary when considering quantity discounts. We differentiate two cases:

Case 1.
Quantity discount is only given for the parts of orders whose quantities exceed q_0.

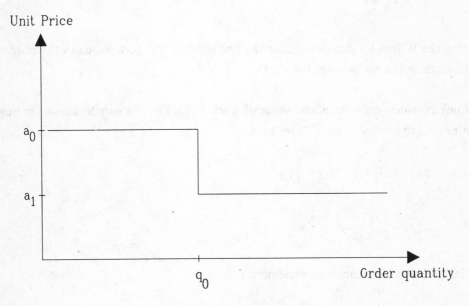

Figure 9.1: Discount scale

$D^* \geq q_0$ is not an interesting case. We, therefore, assume $D^* \leq q_0$ first and ask what the optimal order quantity $\hat{D}$ is if more than q_0 is ordered.

The average unit cost is

$$\bar{C} = \frac{k + q_0 a_0 + (D-q_0)a_1 + h \cdot \frac{D}{2} \cdot \frac{D}{\lambda}}{D}$$

$$= \frac{K}{D} + \frac{h}{2\lambda}D + a_1 , \qquad D \geq q_0 , \qquad (9.1)$$

where $K = k + q_0(a_0 - a_1)$. $\bar{C}$ is convex. We disregard for a moment the condition $D \geq q_0$ and obtain using $d\bar{C}/dD \overset{!}{=} 0$ as minimizing lot size

$$\hat{D} = \sqrt{\frac{2K\lambda}{h}} .$$

It is to be tested whether $\hat{D} > q_0$. For $\hat{D} \leq q_0$, $\bar{C}(D^*)$, $\bar{C}(D)$ is the global minimum. For $\hat{D} > q_0$, $\bar{C}(\hat{D})$ and $\bar{C}(D^*)$ are two relative minima and it remains to be determined which of the two is the global minimum.

A comparison of unit costs results into (note: $\bar{C} = \bar{c} + a$)

$$\sqrt{\frac{2kh}{\lambda}} + a_0 \quad \lessgtr \quad \sqrt{\frac{2Kh}{\lambda}} + a_1 . \qquad (9.2)$$

$\text{Case } D^* \qquad \text{Case } \hat{D}$

Example:

Let $k = 8$, $h = 0.01$, $\lambda = 1$, $q_0 = 100$, $D^* = 40$.

How large must the quantity discount $x = a_0 - a_1$ be so that it pays to take advantage of it?

The advantage is cancelled at

$$0.4 + x = \sqrt{\frac{2(8 + 100x) \cdot 0.01}{1}}$$

$$\Rightarrow x = 1.2.$$

Is it also true that $\hat{D} > q_0$?

$$\hat{D} = \sqrt{\frac{2(8 + 120)}{0.01}} = 160 > q_0.$$

Case 2.

The lower price a_1 is chosen for the whole order quantity D as soon as $D \geq q_0$.
Again, let $D^* < q_0$. Then it is clear that an order quantity $D > q_0$ is not worthwhile.
But maybe $D = q_0$? The cost comparison is again the criterion

$$\sqrt{\frac{2kh}{\lambda}} + a_0 \lessgtr \frac{k}{q_0} + \frac{h}{2\lambda} \cdot q_0 + a_1 . \qquad (9.3)$$

Case D^* \qquad Case $D = q_0$

Example:

With the same cost figures as before, the criterion (9.3) gives

$$0.4 + x \lessgtr \frac{8}{100} + \frac{0.01}{2} \cdot 100 .$$

Indifference is reached at $x = 0.18$. The discount jump is now considerably lower than the first case.

§10   COLLECTIVE OR SINGLE ORDER ?

If a number of goods come from the same supplier, then a collective order can pay off under certain circumstances. Let

k_i, h_i, λ_i :   Fixed order costs, holding cost and rate of demand of good i

k_o :   Fixed order costs for collective order

c_c: Collective order costs per unit time

c_s: Single order costs per unit time

Single Order:
The cost per unit time (without proportional order costs) amounts on the average (compare (3.2)) to

$$c_s = \sum_i \sqrt{2 k_i h_i \lambda_i} \; . \tag{10.1}$$

Collective Order:
The reorder time must be the same for all goods. For an order cycle of length t, each lot requires $D_i = \lambda_i t$. The cost of one cycle per unit time (again without proportional order costs) is (compare (3.1))

$$c_c = \frac{k_o + \frac{1}{2} \sum h_i D_i t}{t} \; . \tag{10.2}$$

Therefore, the objective function is

$$\frac{k_o}{t} + \sum_i \frac{h_i \lambda_i t}{2} \underset{t}{\to} \text{Min}$$

$$\Rightarrow \boxed{\quad T_{opt} = \sqrt{\frac{2k_o}{\sum_i h_i \lambda_i}} \quad} \quad .$$

We substitute the optimal cycle length T in (10.2) and obtain as minimum cost c_c for the collective order

$$c_c = \sqrt{\frac{k_o}{2} \sum_i h_i \lambda_i} + \sum_i \sqrt{\frac{k_o}{2} \frac{h_i^2 \lambda_i^2}{\sum_j h_j \lambda_j}} =$$

$$= \sqrt{\frac{k_o}{2}} \sqrt{\sum_i h_i \lambda_i} + \sqrt{\frac{k_o}{2}} \sum_i h_i \lambda_i \sqrt{\frac{1}{\sum_j h_j \lambda_j}} =$$

$$= \sqrt{\frac{k_o}{2}} \left\{ \sqrt{\sum_i h_i \lambda_i} + \sqrt{\frac{\left[\sum_i h_i \lambda_i\right]^2}{\sum_j h_j \lambda_j}} \right\} \quad , \text{i.e.}$$

$$c_c = \sqrt{2k_o \sum_i h_i \lambda_i} \quad .$$

Compare: $c_s \lessgtr c_s$?

By cost comparison one can cancel the common factors such that the question becomes

$$\boxed{\sum_i \sqrt{k_i h_i \lambda_i} \;\overset{?}{\lessgtr}\; \sqrt{k_0 \sum_i h_i \lambda_i}} \quad .$$

Case 1.

$k_0 = \sum k_i$, i.e. with fixed order costs, a collective order does not have an advantage over a single order. The computation shows

$$\sum_i \underbrace{\sqrt{k_i}}_{=:\,K_i} \cdot \underbrace{\sqrt{h_i \lambda i}}_{=:\,A_i} \;\overset{?}{\lessgtr}\; \sqrt{\sum_i k_i} \sqrt{\sum_i h_i \lambda_i}$$

$$\sum_i K_i \cdot A_i \;\overset{?}{\lessgtr}\; \sqrt{\sum_i K_i^2} \sqrt{\sum_i A_i^2} \quad .$$

The scalar product of both vectors K, A are on the left side and the product of these quantities on the right. Therefore,

$$\sum_i K_i A_i \;(\overset{<}{=})\; \sqrt{\sum_i K_i^2} \sqrt{\sum_i A_i^2} \quad .$$

To avoid trivial cases, we can, as a rule, assume K, A > 0. Then the equality sign disappears and the single order is better than the collective order. Reason: In a single order, the different individual optimal lots D_i^* are ordered. This is not possible in a collective order.

Case 2.

$k_0 = k_i = k$. In this case one expects a cost advantage from the collective order. The computation confirms this. It is

$$\sum \sqrt{h_i \lambda} \overset{?}{\underset{<}{\geq}} \sqrt{\sum h_i \lambda_i}$$

$$\sum A_i \overset{?}{\underset{<}{\geq}} \sqrt{\sum A_i^2} \, .$$

Squaring both sides gives the obvious statement

$$\left(\sum A_i \right)^2 > \sum A_i^2 \quad \text{for } A_i > 0 \, .$$

Case 3.

$k_i = k + \eta_i; \ k_0 = k + \sum_i \eta_i$; i.e., the fixed order costs consist of a basic value k and a product–dependent value η_i . In this case, it could be $c_s < c_c$, as well as, $c_s > c_c$.

§11   OPTIMAL STOCKING IN SERIAL PRODUCTION

Lot sizes occur not only in commercial warehouses but also in production. We consider the simple case of production with continuous withdrawals from the finished parts inventory.

An example is engine production in an automobile factory. The production program for the next half–year has scheduled the assembly of four–cylinder vehicles at a constant rate. These vehicles are manufactured daily on the assembly line. How large is the production lot of these engines?

Let

μ: Production rate, $\mu > \lambda$

D: Lot size minus the continuous withdrawal during production of a lot (net lot size, stocking lot size)

The stock movement has the following characteristic

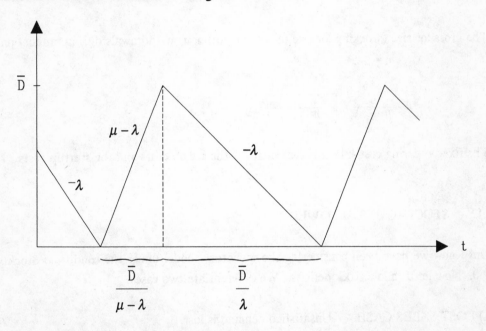

Figure 11.1: Stock control with replenishment through production

The cost per unit time is

$$c = \frac{k + h \cdot \frac{\overline{D}}{2} \cdot \overline{D}(\frac{1}{\lambda} + \frac{1}{\mu - \lambda})}{\overline{D}(\frac{1}{\lambda} + \frac{1}{\mu - \lambda})} \to \underset{\overline{D}}{\text{Min}} \ . \tag{11.1}$$

The optimal net lot size is

$$\overline{D} = \sqrt{\frac{2k}{h} \cdot \frac{1}{\frac{1}{\lambda} + \frac{1}{\mu - \lambda}}} \cdot \qquad\qquad (11.2)$$

Instead of the rate λ in (2.2), the harmonic mean of λ and $\mu - \lambda$ now appears:

$$1 / (\frac{1}{\lambda} + \frac{1}{\mu - \lambda}) \cdot$$

The gross lot size (stocking lot size plus the continuous withdrawals during production) is

$$D = \mu \cdot \frac{\overline{D}}{\mu - \lambda} = \sqrt{\frac{2k\lambda}{h} \cdot \frac{\mu}{\mu - \lambda}} \cdot$$

The fixed ordering cost k in this case includes the set up costs and the startup costs.

§12   STOCK–OUTS ALLOWED

Until now we have been considering the inventory model under the condition: Stock y $\geq$ 0. Now let us allow stock deficits. We differentiate two cases:

a) LOST SALES CASE.   Unsatisfied demand is lost.
b) BACKORDER CASE.   Unsatisfied demand is deferred until supply is again available.

We consider the BACKORDER CASE. In practice, under conditions of certainty, this can only be done by a firm without competition, i.e., a monopolist (with the attitude "the public be damned"). It is, however, different with stochastic demand. In this case, one can not in all cases, even with the best intentions, guarantee a 100% delivery.

As a rule, it will cost something if a shortage occurs. On the other hand, if these costs are not too high, stock deficits may also be cost effective.

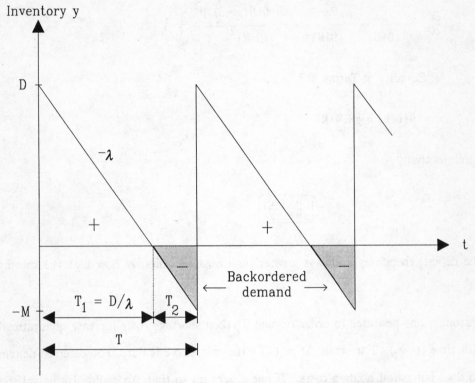

Figure 12.1: Operational characteristic of inventory control with shortages (Backorder case)

The stock deficit M is evaluated with the proportional shortage cost g.

The cost per period is

$$c = \frac{k + h \cdot \dfrac{D}{\lambda} \cdot \dfrac{D}{2} + g \cdot \dfrac{M}{\lambda} \cdot \dfrac{M}{2}}{(D + M)/\lambda} \to \underset{D,M}{\text{Min}} \ . \tag{12.1}$$

$c(D,M)$ is convex. From

$$\frac{\partial c}{\partial M} : -\frac{k\lambda}{(D+M)^2} - \frac{\frac{h}{2} D^2}{(D+M)^2} + \frac{gM(D+M) - \frac{gM^2}{2}}{(D+M)^2} \overset{!}{=} 0$$

$$\frac{\partial c}{\partial D} : -\frac{k\lambda}{(D+M)^2} - \frac{\frac{gM^2}{2}}{(D+M)^2} + \frac{hD(D+M) - \frac{h}{2}D^2}{(D+M)^2} \overset{!}{=} 0$$

$\Rightarrow$ Equality of Terms

$\Rightarrow$ $hD(D+M) = gM(D+M)$

it follows that

$$\boxed{\frac{D}{M} = \frac{g}{h}} \ .$$

(12.2)

The deficit, therefore, is always greater than zero regardless of how high the shortage cost is.

Reason: If one hesitates to order beyond T_i then shortage costs increase quadratically with time $(t-T_i)$. For small $\Delta t = t-T_1$ the cost curve is flat. The deferred demand Δq does not entail holding costs. If one stocks up so that Δq could also be satisfied then one must stock Δq for the whole period T_i.

From (12.2),

$$T_1 = \frac{g}{h+g}\,T;$$

$$T_2 = \frac{h}{g+h}\,T \ .$$

Hence, the cost function becomes

$$c = \frac{1}{T}\,[k + h\,\frac{\lambda T_1}{2}\,T_1 + g\,\frac{\lambda T_2}{2}\,T_2] =$$

$$= \frac{1}{T}\,[k + h\,\frac{\lambda}{2}\,(\frac{g}{g+h})^2 T^2 + g\,\frac{\lambda}{2}\,(\frac{h}{g+h})^2 T^2]\ .$$

$$c \to \underset{T}{\text{Min}} \Leftrightarrow \frac{dc}{dT} = -\frac{k}{T^2} + h\frac{\lambda}{2}\left(\frac{g}{g+h}\right)^2 + g\frac{\lambda}{2}\left(\frac{h}{g+h}\right)^2 \overset{!}{=} 0$$

$$\Rightarrow \quad \boxed{T = \sqrt{\frac{2k}{\lambda}\left(\frac{1}{h}+\frac{1}{g}\right)}} \quad . \tag{12.3}$$

The optimal order is

$$\boxed{D + M = \sqrt{2k\lambda\left(\frac{1}{h}+\frac{1}{g}\right)}} \quad . \tag{12.4}$$

Both these results can be reduced to the formulas (2.2) and (2.3) for $g \to \infty$.

§13 DISCRETE LOT SIZES

In our previous inventory model the order quantity took on a real value. For small lots, however, the requirement of discreteness cannot be ignored. The holding cost per cycle is now

$$\frac{1}{\lambda} \cdot h \sum_{i=0}^{D-1}(D-i) = \frac{h}{\lambda}\sum_{j=1}^{D} j = \frac{h}{\lambda}\frac{D(D+1)}{2} \quad .$$

The period is $\frac{1}{\lambda}$, during which the inventory remains at its present state, i.e., the time between two demands.

The objective function c (cost of an order cycle per unit time, without proportional order cost) is

$$c = \frac{\lambda k}{D} + \frac{h}{2}(D+1) \to \underset{D \in \mathbb{N}}{\text{Min}} \quad . \tag{13.1}$$

The condition for the minimum of the convex function c_n for whole numbers n is

$$c_{n^*} = \underset{n \,\epsilon\, \mathbb{N}}{\text{Min}} \{c_n\} \Rightarrow c_{n^*-1} \geq c_{n^*} \leq c_{n^*+1} \;.$$

(see the following figure 13.1)

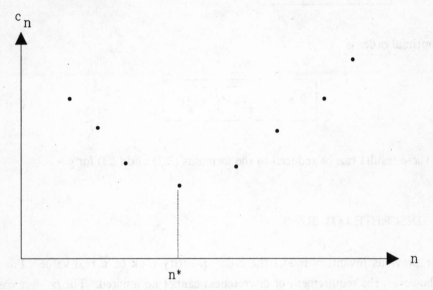

Figure 13.1: Convex function c_n with discrete argument

One considers the first difference

$$\Delta_n: = c_n - c_{n-1} \;.$$

For n* it changes from negative to positive.

Example:
For $\lambda = 1$, $h = 1$, $k = 1$, the optimal lot size D according to the WILSON formula
(2.1) would have been $D = \sqrt{2}$.

Should one round up or down? It is better to proceed from D^* instead of D using the
first differences.

We have $c(0) >> 0$. For the objective function (13.1) it is

$$\Delta_1 = \frac{1}{1} + 1 - c(0) < 0;$$

$$\Delta_2 = \frac{1}{2} + \frac{3}{2} - 2 = 0; \ \Rightarrow \ \text{the Minimum occurs in two places } n = 1 \text{ and } n = 2$$

$$\Delta_3 = \frac{1}{3} + 2 - 2 > 0 \ .$$

Hence $D^* = 1$ and $D^* = 2$ are both correct solutions.

§14 CONSIDERATION OF SHELF SPACE IN INVENTORY

<u>Reserved Shelf Space</u>
In a multi–product inventory, we assume a specific shelf space is always designated for an item in order to have quick access to this particular item. Hence, the holding cost depends on the size of the reserved area. It is similar to the maximum stock quantity, D.

Let

h_1 : holding cost proportional to quantity

h_2 : holding cost proportional to shelf space

The cost per unit time for an inventory cycle of length D/λ is

$$c = \frac{k + h_1 \frac{D}{2} \cdot \frac{D}{\lambda} + h_2 D \cdot \frac{D}{\lambda}}{D/\lambda} \ .$$

It is minimized when the lot size is

$$\boxed{D = \sqrt{\frac{2k\lambda}{h_1 + 2h_2}}} \ . \tag{14.1}$$

Hence, the effective inventory cost is the sum of the holding cost h_1 and twice the cost of the reserved storage space.

Periodic Determination of Order Quantities

Two products are stored. Shelf space is not reserved. The order pattern is the same for both products. The cycle length is T. One can shift the order points of the two products such that the maximum required total storage area is minimized. We denote

τ :   Phase displacement of orders of product 2.

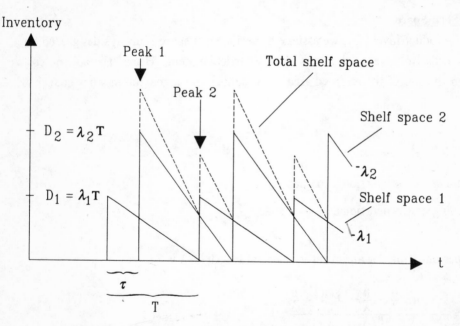

Figure 14.1: Periodic stock movement (single and composite)

Total inventory manifests itself at two peaks.

Peak 1: Order of product 2
Peak 2: Order of product 1

The optimal phase displacement results from the condition

$$\underset{\tau}{\text{Min}} \{\text{Max} \{\text{Peak 1} | \text{Peak 2}\}\} .$$

The minimum is reached at the point where the two peaks have the same height

$$\lambda_2 T + \lambda_1 (T - \tau) = \lambda_1 T + \lambda_2 \tau \qquad (14.2)$$

$$\boxed{\tau = \frac{\lambda_2}{\lambda_1 + \lambda_2} T} \quad . \qquad (14.3)$$

We substitute τ in (12.2) and obtain the maximum stock

$$\boxed{\text{Max} \{y_1 + y_2\} = \frac{\lambda_1^2 + \lambda_1 \lambda_2 + \lambda_2^2}{\lambda_1 + \lambda_2} \cdot T} \quad . \qquad (14.4)$$

It is symmetric in λ and proportional to $T/\lambda_1 + \lambda_2$.

When the total rate $\lambda_1 + \lambda_2$ is constant, the expression

$$\frac{\lambda_1^2 + \lambda_1 \lambda_2 + \lambda_2^2}{\lambda_1 + \lambda_2}$$

assumes a minimum for $\lambda_1 = \lambda_2$. The proof is left to the reader.

§15   BUDGET RESTRICTION

In a multi–product warehouse, the individual goods compete for storage space. In a small storage room, one cannot expect that the total area will cover storage of the optimal lot size D_i from (14.1) for each good i. As a rule one must get by with a fraction of D_i. This leads to an inventory model with capacity restrictions. Instead of limiting the storage area, the available capital could also be limited: either the current inventory asset or the current account, limited by the available credit line, in case all orders within an inventory cycle are paid for at the same time.

Let

b_i :   Space requirement or price per unit of product i

b_o :   Total inventory capacity or budget

x_i :   Lot size

We minimize the costs $\sum\limits_i c_i$ of a cycle per period (compare (2.1))

$$\underset{x_1,\ldots,x_N}{\text{Min}} \left\{ \sum_{i=1}^{N} \left[\frac{k_i \lambda_i}{x_i} + \frac{h_i}{2} x_i \right] \right\} \tag{15.1}$$

with the constraint

$$\sum_{i=1}^{N} b_i x_i \le b_o \tag{15.2}$$

Using the method of Lagrange multipliers:

The constraint (15.2) is added to the objective function using the Lagrange multiplier β. The expanded function is called the Lagrange function L

$$L = -\left[\sum_i \frac{k_i \lambda_i}{x_i} + \sum_i \frac{h_i}{2} x_i\right] + \beta\underbrace{\left[b_0 - \sum_i b_i x_i\right]}_{!} . \tag{15.3}$$

$\uparrow$
since Min! $\qquad \geq 0$

L is a concave function, therefore it is sufficient for an extremum

$$\frac{dL}{dx_i} \overset{!}{=} 0: \frac{k_i \lambda_i}{x_i^2} - \frac{h_i}{2} - \beta b_i = 0,$$

$$\boxed{x_i = \sqrt{\frac{k_i \lambda_i}{\frac{h_i}{2} + \beta b_i}}} . \tag{15.4}$$

For $\beta = 0$, i.e., if the budget restriction is never effective, (15.4) again becomes the old WILSON Formula (2.2). A comparison of both formulas shows that the budget restriction occurs in the form of higher inventory holding costs. If one uses only interest to represent holding costs, at a rate which is profitable given the available capital (e.g. the firm's internal rate of return) and if b_i is the capital investment per unit of product i, then the budget restriction leads to an increase in the nominal interest cost.

If one interprets the constraint (15.2) as a space restriction, then it has the effect of an additional rental space of 2β per unit area.

When does a reduction of order quantity for all goods (and, hence, the inventories) occur in the same proportion? It is sufficient that

$$b_i \sim h_i ,$$

since with $b_i = \alpha h_i$, $\alpha \in \mathbb{R}$, i = 1,2,...,N, (15.4) results in

$$x_i = \sqrt{\frac{2k_i\lambda_i}{h_i(1+2\alpha\beta)}} \; .$$

The optimization problem (15.1), (15.2) can also be formulated as a non–linear program

$$\min_{x_1,\ldots,x_N} \left\{ \sum_{i=1}^{N} \left[\frac{k_i\lambda_i}{x_i} + \frac{h_i}{2} x_i \right] \right\} \; ;$$

such that   1) $\displaystyle\sum_{i=1}^{N} b_i x_i \leq b_0$;

2) $x_i \geq 0, \quad i = 1,2\ldots,N.$

Since the optimization problem is now restricted to $x_i \geq 0$, a boundary extreme could occur, but that is never the case in the above objective function.

Determination of β: $\frac{\partial L}{\partial \beta} \geq 0$ applies. Since all $x_i > 0$, $\frac{\partial L}{\partial \beta} = 0$.

From this follows

$$b_0 = \sum_{i=1}^{N} \sqrt{\frac{\lambda_i k_i b_i^2}{\frac{h_i}{2} + \beta b_i}} \; .$$

The bigger b_0 is, i.e., the weaker the constraint equation, the smaller β becomes.

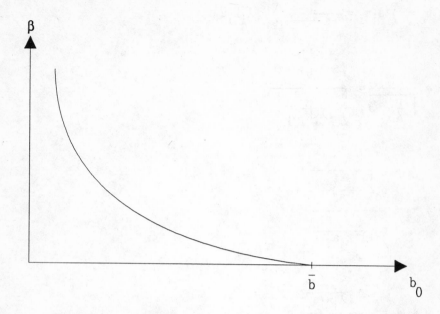

Figure 15.1: Dependence of an additional cost β on the budget

At $b_0 > \bar{b}$, the budget is no longer fully utilized.

Example: Inventory Costs

r	: Interest
$h_i = ra_i$	: Capital budget costs
$b_i = a_i$	: Proportional order costs
$\sum a_i x_i \leq b_0$	: Budget constraint

Then

$$x_i = \sqrt{\dfrac{\lambda_i k_i}{r \dfrac{a_i}{2} + \beta a_i}} = \dfrac{1}{\sqrt{\dfrac{r}{2} + \beta}} \sqrt{\dfrac{\lambda_i k_i}{a_i}} \ .$$

β is the shortage cost which is added to the capital.

$$b_0 = \sum_{i=1}^{N} a_i x_i = \frac{1}{\sqrt{\frac{r}{2} + \beta}} \sum_{i=1}^{N} \sqrt{\lambda_i k_i a_i}$$

$$\Rightarrow \frac{1}{\sqrt{\frac{r}{2} + \beta}} = \frac{b_0}{\sum_{i=1}^{N} \sqrt{\lambda_i k_i a_i}}$$

$$\Rightarrow x_i = \frac{b_0 \sqrt{\frac{\lambda_i k_i}{a_i}}}{\sum_{j=1}^{N} \sqrt{\lambda_j k_j a_j}}$$

$$\Rightarrow a_i x_i = \frac{\sqrt{\lambda_i k_i a_i}}{\sum_{j=1}^{N} \sqrt{\lambda_j k_j a_j}} b_0 \ .$$

From the last equation it is obvious that the relationship of each $a_i x_i / a_j x_j$ to each of the individual lot sizes does not change if one increases the budget b_0.

§16   KNOWN BUT VARYING DEMAND

Until now the demand rate was considered as constant over an infinite time horizon. For commercial warehouses this occurs only in exceptional cases. There is a similar situation with manufacturing inventories, for example, where purchased parts are kept in stock to be used as components in serial production. A constant demand rate is also rare in that case. Hence, we drop this restrictive assumption. Let

λ_i :   Demands during the periods i = 1,2,...n

Demand is, therefore, known up to period n.

Example: Automobile Factory

Depending on customer's preferences, a vehicle can be equipped with a wood grain steering wheel. The daily requirements for the steering wheel are determined from the bill of materials for the vehicles prepared from the preceding quarter. These are ordered from a supplier. What is the optimal lot size?

One can formulate this problem as an integer programming problem with a planning horizon of n. However, this would not be a proper use of the model since, after a fraction of the planning horizon has passed, new information about the requirements for the period after n already exists. Strictly speaking, we are dealing with a rolling planning horizon. It would be a fruitless effort to want to solve this problem exactly.

There are two possibilities for this model:

a)   Either one knows how the problem behaves in the future, i.e, one can, at least, give probabilities for future demands. Such models are treated later.

b)   Or, one considers it as an infinite horizon problem. One asks how long an order will suffice. The time period T is chosen such that the average costs are minimized. This procedure is repeated each time the stock sinks to zero. Surely, this method is not optimal but is more practical than the exact method because one need not wait for a new decision until the whole planning horizon has passed. In the literature, this method is known as the SILVER–MEAL heuristic. (SILVER & MEAL (1973)).

The objective function is to "minimize the average cost of the first order cycle"

$$c(T) \rightarrow \underset{T}{\mathrm{Min}} \qquad (16.1)$$

$$c(T) = \frac{k}{T} + \frac{h}{T}[\lambda_T T + \lambda_{T-1}(T-1) + ... + \lambda_1] \qquad (16.2)$$

Note: The quantity λ_i is stored for i periods. Minimization occurs in the discrete variable T. The optimality conditions are

$$c(T) - c(T - 1) \le 0 ;$$

$$c(T + 1) - c(T) \ge 0 .$$

(16.3)

These are, however, only necessary but not suffficient, since (16.2) is not convex. The convexity is ensured if $\frac{h}{T} \sum\limits_{i=1}^{T} i\lambda_i$ is monotone, i.e., for

$$T(T + 1)\lambda_{T+1} > \sum_{i=1}^{T} i\lambda_i ; \quad T = 1,2, ..., n.$$

Example: Let k = 12, h = 0.1, n=8

T	1	2	3	4	5	6	7	8
λ_T	5	3	6	2	4	3	4	7
k/T	12	6	4	3	2.4	2	1.7	1.5
$\sum i\lambda_i$	5	11	29	37	57	75	103	159
$\frac{h}{T} i\lambda_i$	0.5	0.55	0.97	0.93	1.14	1.25	1.47	1.99
c(T)	12.5	6.55	4.97	3.93	3.54	3.25	3.17	3.49
							T = 7	

for h=1, two local minima are obtained!

c(T)		17	11.5	13.7	12.3	13.8	14.5	16.4	21.4
			T=2						

We now consider the problem in continuous time

$$c(T) = \frac{1}{T}\left[k + h \int_0^T t\lambda(t)dt\right] \quad .$$

Minimizing with respect to T:

$$\frac{dc}{dT} \stackrel{!}{=} 0: \quad -\frac{k}{T^2} + h\,\frac{T^2\lambda(T) - \int_0^T t\lambda(t)dt}{T^2} = 0$$

$$\frac{k}{T} = h\left[T\lambda(T) - \frac{1}{T}\int_0^T t\lambda(t)dt\right]$$

$$\boxed{\frac{1}{T}\left[k + h\int_0^T t\lambda(t)dt\right] = hT\lambda(T) \quad .}$$

Average cost with respect to time for a sufficient order · · · Limit cost of the order cycle

This equation illustrates the economic principle in cost minimization:

The average cost must be equal to the marginal cost.

Other methods, for example, are the method of the smoothed economic lot size (minimization of the unit cost of a lot) and the "Part–Period–Method" of DeMATTEIS and MENDOZA (1968) (Minimization criterion: the order and holding costs are equal for optimal lot sizes). The latter, as a rule, achieves better results (compare OHSE (1970)).

Studies by KNOLMAYER (1985) have shown that the heuristic by SILVER and MEAL gives the best results for rates of demand which fluctuate about a constant average.

Two modifications of this method stem from SILVER and MILTENBURG (1984). One was developed for the case of monotonically decreasing demand and the other for the case of sporadic demand.

Of course, there are some situations in which it is meaningful to solve the lot size problem exactly; for example, if a branch is to be closed in the near future. The production program during this phase—out is fixed and with it the demand for raw materials in the different periods. A rolling plan is not applied here. Dynamic optimization is a possible method of solution for this problem formulated by WAGNER and WHITIN (1958).

In general, the algorithm of WAGNER and WHITIN performs poorly compared to the SILVER—MEAL heuristic (BLACKBURN and MILLEN (1980)). However, CHAND (1982) has modified it such that, according to his own statement, it performs much better than the SILVER—MEAL heuristic.

§17   FIXED DELIVERY PERIOD τ

If the delivery period τ is not zero but positive, constant and known, then the time of ordering and the periods of maximum stock levels should be differentiated. Obviously, the order must be placed τ time periods before the stock becomes empty. The stock level is at its optimal order point when

$$y = \lambda \cdot \tau$$

If the delivery period τ is longer than the duration of a cycle

$$T = \frac{1}{\lambda} D,$$

then orders occur in each time period and, in certain periods, more than once. If that is not allowed, one must always order the quantity $\lambda\tau$ at the moment when the last order has been delivered. Because of this, costs increase compared to the case in which frequent orders are made. In general, firms seek to avoid early deliveries in the same way they attempt to avoid late deliveries by imposing contract penalties, etc. The punctuality and reliability of suppliers in Japan are cited by US automobile companies as production advantages of their Japanese competitors.

Weak Demand

For a product which is rarely demanded, the question is asked whether one should stock this product at all.

a) Do not hold in stock: A penalty cost per turnover is incurred in the amount of $g\tau$
b) Hold in stock: A holding cost per turnover is incurred in the amount of h/λ.

Fixed order costs are ignored for comparison. The product will not be stocked for $g\tau < h/\lambda$, i.e.,

$$\boxed{\lambda\tau < \frac{h}{g}} \; . \qquad (17.1)$$

A similar situation arises in the mail–order business. Each customer placing an order is seen as a marketing outlet. In this extreme form of decentralization, the rate of sales per outlet and per product is low. Savings in terms of inventory holding costs are partly returned as a price advantage to the customer.

This model also occurs in the pharmaceutical industry. The demand for a given medicine at an outlet (pharmacy) is low and the delivery time is very short (a few hours). For this reason, pharmacies stock only a basic assortment of medicines.

§18   SAFETY STOCK WITH STOCHASTIC DELIVERY TIME
(including JUST IN–TIME PRODUCTION)

We continue the discussion of the case of a constant and known demand. Let the delivery time τ now be a random variable with an expected value μ_{τ}. If one specifies the order point $s_1 = \lambda\mu_T$ (this is already the demand during the expected delivery time μ_{τ}), then, in the case of a symmetrical delivery time distribution, one would have as many stocks as shortages immediately before an order arrives. This is only optimal if the inventory cost rate, h, is equally as large as the shortage cost rate, g. By increasing the order point to $s_2 > s_1$, the risk of shortage is reduced. The amount s_2-s_1 is the <u>safety stock</u>. Its purpose is to cover any deviations in expected delivery times.

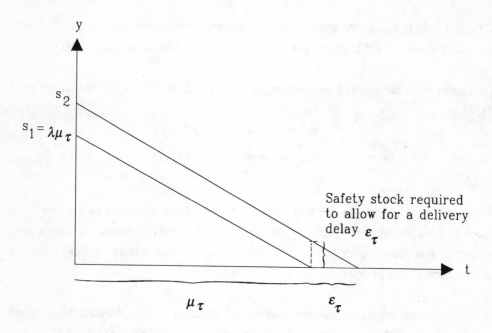

Figure 18.1: Stock movement with and without safety stock

We assume that a delivery time μ_τ was agreed upon with a supplier. Because of unforeseen circumstances, delays (e.g., production bottleneck, slow customs clearance procedures) or early deliveries (e.g. as a result of route planning of deliveries) may occur. The processing of orders during receipt of goods and during quality control may also cause fluctuations. These deviations are unexpected and are, therefore, considered in the model as an occurrence of a random error ϵ_τ.

ϵ_τ :   Random deviation from the agreed–upon delivery schedule, random quantity
         with distribution function $P(\epsilon_\tau)$

The total delivery time τ

$$\tau = \mu_\tau + \epsilon_\tau \qquad\qquad (18.1)$$

is then a random quantity.

The problem of uncertain delivery time is often solved heuristically. One determines a percentage service level

$$\text{SERVICE LEVEL } \beta = \frac{E\{\text{Satisfied Demand}\}}{E\{\text{Total Demand}\}} \times 100,$$

e.g. $\beta = 97\%$. The probability of shortage is

$$\text{Prob}(y < 0) = 1 - \beta/100. \qquad\qquad (18.2)$$

The safety stock $s_2 - s_1$ should be large enough to allow the firm to achieve the given service level and the desired shortage probability.

If one knows the distribution function $P(\epsilon_\tau)$ of the error variable, then the safety stock may be derived from (18.2). The value $s_1 = \lambda\mu_\tau$ is known. Shortages occur if the delivery time is longer than the duration of the safety stock

$$\mu_\tau + \epsilon_\tau > \frac{s_2}{\lambda} ,$$

i.e.

$$\epsilon_\tau > \frac{s_2}{\lambda} - \mu_\tau.$$

For a fixed s_2, the probability of the occurrence of shortages is

$$\text{Prob}(y < 0) = \text{Prob}(\epsilon_T > \frac{s_2}{\lambda} - \mu_T), \qquad (18.3)$$

$$= 1 - P(\frac{s_2}{\lambda} - \mu_T).$$

We choose s_2 such that the probability of shortage takes a desired valued, i.e., it satisfies (18.2). Hence it must follow

$$1 - \beta/100 = 1 - P(\frac{s_2}{\lambda} - \mu_T),$$

or

$$\beta/100 = P(\frac{s_2}{\lambda} - \mu_T). \qquad (18.4)$$

One obtains the order point s_2 from the β–percentile of the distribution function $P(\epsilon_T)$. If, for example, the error variable is normally distributed $N(0, \sigma_T)$, then from (18.4)

$$\beta/100 = N(\frac{s_2}{\lambda} - \mu_T)$$

and by standardizing to the $N(0,1)$ normal distribution

$$\beta/100 = N_{(0,1)}(\frac{s_2}{\lambda \sigma_T} - \frac{\mu_T}{\sigma_T}).$$

Let $\epsilon_{\beta\%}$ be the β–percentile of the $N(0,1)$ distribution. For s_2, it results into the condition

$$\epsilon_{\beta\%} = \frac{s_2}{\lambda\sigma_\tau} - \frac{\mu_\tau}{\sigma_\tau}$$

$$\Rightarrow \quad \boxed{s_2 = \underbrace{\epsilon_{\beta\%}\,\lambda\sigma_\tau}_{SB} + \underbrace{\lambda_{\mu\tau}}_{s_1}} \quad . \qquad (18.5)$$

Since $\lambda\mu_\tau = s_1$, we obtain for the safety level SB the term

$$\boxed{SB = \epsilon_{\beta\%}\,\lambda\sigma_\tau.} \qquad (18.6)$$

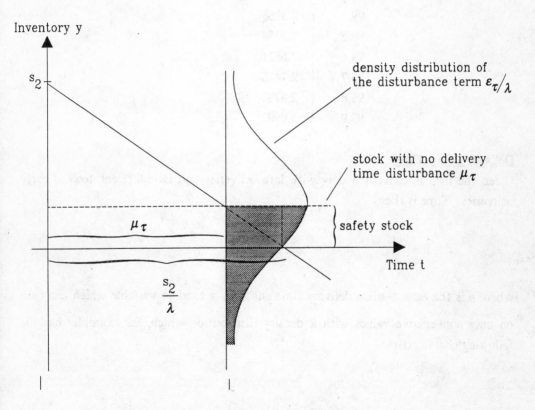

Figure 18.2: Safety stock level and shortage probability with order point s_2.
The density function ϵ_τ is superimposed on the graph.

The corresponding percentages with $(0,1)$–normally distributed delivery times for some service levels are given in Table 18.1.

Table 18.1

Service Level β	$\epsilon_\beta\%$
90	1.2816
95	1.6449
96	1.7507
97	1.8808
98	2.0537
99	2.3263
99.5	2.5758
99.6	2.6521
99.7	2.7478
99.8	2.8782
99.9	3.0902

Delivery Delays Only

Often the firm is concerned only with late deliveries and is indifferent toward early deliveries. Time is then

$$\tau = \mu + \epsilon_\tau,$$

where μ is the agreed–upon delivery time and ϵ_τ is a random variable which can take on only nonnegative values with a density distribution which, for example, has the following characteristic

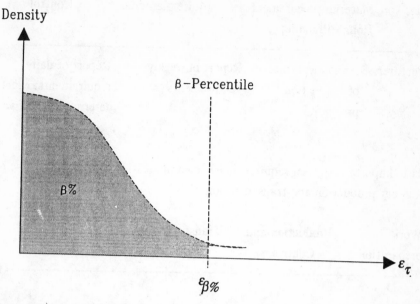

Figure 18.3: Density distribution of a random delivery delay ϵ_τ

If one wants to ensure himself against $\beta\%$ of all delays, then one has to store a safety stock which is enough for the period until the β–percentile, i.e., until $\epsilon_\beta\%$

$$\boxed{SB = \lambda\epsilon_{\beta\%}} \tag{18.7}$$

Note: In comparison to (18.6), the distribution of the delivery delay here is not normalized. Hence the term ϵ_τ does appear as a multiplier.

Safety Stock in Just–in–Time Production

We now consider as products component parts which are manufactured by a supplier and put together on an assembly line. If economic or space limitation reasons dictate, the component parts may be produced on the day they are needed and delivered to the assembly line "just–in–time". The time planning runs, in principle, according to the following schema:

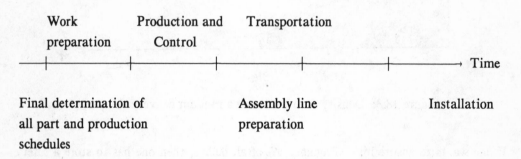

Supplier: Material procurement   Finite planning   Control
              Rough Planning
——→ Time
Manufacturer: Announcement   Report of weekly   Report of daily
              of production   requirements   requirements; exact
              program   determination of variants.

The last important phase requires close communication and strong discipline in carrying–out production and transportation

Work   Production and   Transportation
preparation   Control
——+————————————+——————————+——————————+————————+—————→ Time

Final determination of   Assembly line   Installation
all part and production   preparation
schedules

Nevertheless, delays in deliveries may occur. These must be buffered with a safety stock at the assembly line. It is calculated as before according to the formula (18.7). Note, however, that contrary to inventory models, the schedules are not stock–dependent but are directed by the production flow of the manufacturer. Hence, one needs a time buffer instead of a quantity buffer. Instead of the safety stock, an order is placed earlier within the interval $\epsilon_\beta \%$. Quite naturally, the consequence of this early order is also an early delivery and, hence, a stock at the assembly line.

A number of models implemented in practice works with a given service level which is held to be cost minimizing. In OR Literature, one also finds a number of complicated stochastic service level models, e.g., by H. SCHNEIDER, CH. SCHNEEWEISS, J. ALSCHER and M. KÜHN (see ALSCHER, KÜHN and SCHNEEWEISS (1986) and the references given there).

The service level must, however, be chosen carefully because it influences the expected total costs. Fixing the service level at a specific value must strictly be a result of optimization (what is the optimal service level?). In fact, the service level in an inventory problem, in which the costs are to be minimized, depends on how expensive the inventory shortages are. One shoudl,therefore, work directly with the shortage costs instead of the given service level. This does restrict generality since the service level and shortage costs are equivalent. A given shortage cost is assigned a specific service level in optimal lot sizes and vice versa.

In the following chapters, stochastic inventory models are discussed in which shortage costs are considered. Inventory managers sometimes evade the issue of shortage costs either by using rough approximations or by ignoring them altogether and setting arbitrary service levels. The first priority of inventory managers in this instance should be to attempt to determine shortage costs accurately. At times, however, it is not possible to determine exactly the shortage costs. In these cases, the determination of the service level is also as arbitrary as the arbitrary determination of the shortage costs. In these situations, it is more meaningful to vary the service level or shortage costs as a simulation parameter and to specify a concrete value as feasible from the results.

CHAPTER 2:
THE WILSON MODEL WITH POISSON DEMAND

§19   POISSON PROCESS

The POISSON PROCESS will be introduced as a preparation for the discussion of inventory models with stochastic demand. It goes back to BORTKIEWITZ who studied the number of officers in the Prussian Army who were killed by horse kicks.

We derive the Poisson Process using an example of a demand for spare parts. Demand always occurs when a defective part is to be replaced.

A single Part:
Probability of a defect during Δt : $p \cdot \Delta t$;
Probability of no defects during Δt: $1 - p \cdot \Delta t$.

n Parts:
Under the assumptions:
a) the parts do not influence each other; and,
b) the probability of p defects is the same for each part

the probability of u defects in Δt is: $\binom{n}{u}(p\Delta t)^{u}(1 - p\Delta t)^{n-u}$.

Very many parts:
Assumption:
$n \to \infty$, $p \to 0$, but $n \cdot p$ is finite: $n \cdot p = \lambda =$ constant
Probability of no defects in Δt: $p_0(\Delta t)$

$$p_0(\Delta t) = \lim_{n \to \infty} \binom{n}{0}(p\Delta t)^{0}(1 - p\Delta t)^{n}$$

$$= \lim_{n \to \infty} (1 - \frac{\lambda}{n} \cdot \Delta t)^{n} = e^{-\lambda \Delta t} .$$

Observation of defects over time:

Assumptions:

a) infinite no. of parts

b) the probability that any part is defective during the period Δt is $\lambda \Delta t$

c) the probability that many parts are defective in Δt ($\Delta t << \epsilon$), is small and can be ignored,

d) individual defects are independent of each other.

We already know that

$$p_0(t) = e^{-\lambda t} \qquad (19.1)$$

and now compute the probability of u losses in $[0, t]$, $u = 1,2,...$:

Position of Δt at anytime $[0, t]$

$$P_1(t) = \int_0^t e^{-\lambda \tau} \lambda e^{-\lambda(t-\tau)} d\tau = \int_0^t \lambda e^{-\lambda t} d\tau = \lambda t e^{-\lambda t}$$

$$p_2(t) = \int_0^t p_0(\tau) \lambda p_1(t - \tau) d\tau =$$

$$= \int_0^t e^{-\lambda \tau} \lambda \cdot \lambda(t - \tau) e^{-\lambda(t-\tau)} d\tau = \frac{(\lambda t)^2}{2} e^{-\lambda t}$$

$$\vdots$$

in general, u defects in $[0,t]$:

$$\boxed{p_u(t) = \frac{(\lambda t)^u}{u!} e^{-\lambda t}} \qquad , \; u \in \mathbb{N}_0 . \tag{19.2}$$

POISSON PROCESS

Hence, the occurrence of a demand according to a Poisson process (the so-called Poisson demand) is described as the demand for each single component where the time between two successive demands is exponentially distributed (19.1). As before, we have

$\lambda :$ Constant demand rate

Important Property of the Poisson Demand

Constant demand means that the probability of an instantaneous demand is independent of the time elapsed since the occurrence of the last demand. This "lack of memory" is unique for continuous time considerations. There is no other continuous demand type with this characteristic. Lack of memory means

$$p_0(t + \Delta t) = p_0(t) p_0(\Delta t). \tag{19.3}$$

This is the defining functional equation of the exponential function.

Application:

It can be seen clearly that the assumption of a Poisson demand is relevant when there is a very large customer base whereby each customer orders sporadically and independent of the others. The so-called "lumpy demand" falls under this situation.

Poisson Distribution:

If demand is described by the Poisson process, then the demand occurring at period $t = 1$ is Poisson distributed

$$\boxed{p_u = \frac{\lambda^u}{u!} e^{-\lambda}} \qquad , u \in \mathbb{N}_0.$$ (19.4)

The Poisson distribution belongs to the so-called FAMILY OF BINOMIAL DISTRIBUTIONS. This family includes

the Bernoulli Distribution: $\qquad\qquad p_0 = 1 - p;\ p_1 = p;\ 0 < p < 1;$

the Binomial Distribution: $\qquad\qquad p_{u,n} = \binom{n}{u} p^u (1-p)^{n-u}\ ;$

the Negative Binomial Distribution: $\qquad p_{u,n} = \binom{-n}{u}(-p)^u(1-p)^n\ ;$

the Geometric Distribution: $\qquad\qquad p_u = (1-p)p^u\ .$

To conveniently compute the expected value μ and variance σ^2, we use the method of the generating function.

Method of the Generating Function

Let $p_1,\ p_2,\ ...$ be a discrete distribution of an integer-valued random variable u. The function

$$G(x) = \sum_{u=o}^{\infty} p_u x^u$$

is called the GENERATING FUNCTION. It is

$$G'(x)\big|_{x=1} = \sum u p_u = m_1 = \mu \qquad\qquad \text{Expected Value}$$

$$\boxed{G'(x)\big|_{x=1} = \mu}$$

(19.5)

$$G''(x)\big|_{x=1} = \sum u^2 p_u - \sum u p_u = m_2 - m_1.$$

Since $\sigma^2 = m_2 - m_1^2$, one obtains the variance σ^2

$$\boxed{\sigma^2 = G''(x)\big|_{x=1} + G'(x)\big|_{x=1} - \left[G'(x)\big|_{x=1}\right]^2}\ .$$

(19.6)

For the Poisson Distribution,

$$G(x) = \sum_{u=0}^{\infty} \frac{\lambda^u}{u!} e^{-\lambda} x^u = e^{\lambda(x-1)} \qquad \Rightarrow \boxed{\mu = \lambda} \quad \boxed{\sigma^2 = \lambda}$$

For the Poisson Process,

$$G(x,t) = \sum_{u=0}^{\infty} \frac{(\lambda t)^u}{u!} e^{-\lambda t} x^u = e^{\lambda t(x-1)} \qquad \Rightarrow \boxed{\mu_t = \lambda t}$$

$$\boxed{\sigma_t^2 = \lambda t}$$

For the Binomial Distribution,

$$G(x) = (1 - p + px)^n \quad \Rightarrow \boxed{\mu = np} \quad \boxed{\sigma^2 = np(1-p)}\ .$$

For the Negative Binomial Distribution

$$G(x) = \left(\frac{1-p}{1-px}\right)^k \quad \Rightarrow \boxed{\mu = \frac{np}{1-p}} \quad \boxed{\sigma^2 = \frac{np}{1-p}\left(1 + \frac{p}{1-p}\right)}\ .$$

For the Geometric Distribution

$$G(x) = \frac{1 - p}{1 - px} \quad \Rightarrow \quad \boxed{\mu = \frac{p}{1 - p} \quad \sigma^2 = \frac{p}{(1 - p)^2}} \ .$$

The generating function has the following characteristics:

a)   A distribution φ corresponds 1:1 to its generating function

b)   The generating function of the convolution of distributions $\varphi 1$, $\varphi 2$ of two independent random variables is the product of the generating functions of each of the distributions

$$\varphi_1 \otimes \varphi_2 \ \to \ G_{\varphi_1}(x) \cdot G_{\varphi_2}(x)$$

c)   Let $v = f(w)$, w is a random variable with distribution φ_w, then the generating function of the distribution φ_v is the function $G_v(x)$

$$G_{\varphi_v}(x) = G_{\varphi_w}(f(x))$$

With these characteristics, the relationships within the family of binomial distributions can be shown. We will use the last property to develop the first two moments of the so–called stuttering Poisson process.

Compound Poisson Process

An event described by the Poisson process means "demand for one part". Under a compound Poisson process, many parts (or none) may be demanded per event. The time between two events is, as before, exponentially distributed. The number of required parts per event also obeys a distribution.

Example:

A beer salesman sells door–to–door by talking with prospective clients. The duration of each sales talk is exponentially distributed. We consider the end of a sales visit as an event. As a result, u cases of beer may be sold, $u = 0, 1, 2, \ldots$. Let

w_u :   Probability that each sales talk will result in u cases of beer being sold.

Case 1:
Let w_u be Bernouilli distributed, i.e.,

w_1 :   Talk is successful, sale of a case of beer

w_0 :   Talk is a failure, no sale

$$w_1 = 1 - w_0 =: w$$

$w_u(n)$:   Probability of u successes out of n sales talks

$$w_u^{(n)} = (1 - w)w_u^{(n-1)} + w \cdot w_{u-1}^{(n-1)}$$

$$w_u^{(n)} = \binom{n}{u}w^u(1 - w)^{n-u} \qquad \text{Binomial Distribution}$$

$p_u(t)$:   Probability, that u cases are sold up to period t

$$p_u(t) = \sum_{n=u}^{\infty} w_u^{(n)} \frac{(\lambda t)^n}{n!} e^{-\lambda t}$$

$\uparrow$
since at least u talks are needed

$$= \sum_{n=u}^{\infty} e^{-\lambda \cdot t} \frac{(\lambda t)^n}{n!} \binom{n}{u}w^u(1 - w)^{n-u}$$

$$= \frac{e^{-\lambda t}w^u(\lambda t)^u}{u!} \underbrace{\sum_{n=u}^{\infty} \frac{(1 - w)^{n-u}}{(n - u)!} (\lambda t)^{n-u}}_{= e^{(1 - w)\lambda t}}$$

$$P_u(t) = \frac{(w\lambda t)^u}{u!} \, e^{-w\lambda t} \qquad , u \in \mathbb{N}_0 \; . \qquad (19.7)$$

These probabilities describe a compound POISSON PROCESS (with Bernoulli distribution).

The generating function is

$$
\begin{aligned}
G(x,t)_{P.Bern.} &= G_P.(G_{Bern.}(x),t) \\
&= e^{\lambda t[G_{Bern.}(x) - 1]} \\
&= e^{\lambda t[1 - w + wx - 1]} \\
&= e^{w\lambda t[x - 1]} \qquad \Rightarrow \qquad \boxed{\mu_t = \sigma_t^2 = \lambda w t} \; .
\end{aligned}
$$

The corresponding distribution is the compound POISSON DISTRIBUTION (with Bernoulli's distribution)

$$\boxed{P_u = \frac{(w\lambda)^u}{u!} \, e^{-w\lambda}} \qquad\qquad \boxed{\mu = \sigma^2 = \lambda w} \; .$$

Case 2:

Let w_u be geometrically distributed:

$$w_u = (1-w)w^u, \quad 0 < w < 1.$$

The compound process is called the stuttering Poisson process, $w_u^{(n)}$, i.e, we compute the probability that u cases of beer will be sold in n sales talks over the generating function

$$G(x)_{geom.} = \frac{1-w}{1-wx} \; .$$

The generating function of the n–fold convolution of the geometric distribution is

$$[G(x)_{geom.}]^n = [\frac{1 - w}{1 - wx}]^n \quad .$$

If one compares this expression with the negative binomial distribution, one sees that the n–fold convolution of the geometric distribution results in a negative binomial distribution with the power –n

$$NB: (1-z)^{-n} = 1 + \frac{nz}{1!} + \frac{n(n+1)z^2}{2!} + \ldots + \frac{(n+u-1)!z^u}{(n-1)!u!} + \ldots$$

$$= \sum_{u=0}^{\infty} \binom{n+u-1}{u} z^u \quad .$$

Hence,

$$(1-wx)^{-n} = \sum_{u=0}^{\infty} \binom{n+u-1}{u} w^u x^u \quad ,$$

and, therefore

$$[G(x)_{geom.}]^n = \sum_{u=0}^{\infty} \underbrace{\binom{n+u-1}{u}(1-w)^n w^u}_{= w_u^{(n)}} x^u \quad .$$

$$p_u(t) = \sum_{n=u}^{\infty} \frac{(\lambda t)^n}{n!} e^{-\lambda t} \binom{n+u-1}{u} (1-w)^n w^u$$

$$\boxed{p_u(t) = \frac{e^{-\lambda t} w^u (1-w)\lambda t}{u!} \sum_{n=u}^{\infty} \frac{[\lambda t(1-w)]^{n-1}}{(n-1)!n!} (n+u-1)!} \quad . \qquad (19.8)$$

STUTTERING POISSON PROCESS
(with Geometric Disrtibution)

The generating function is

$$G(x,t)_{\substack{st.P. \\ geom.}} = e^{\lambda wt(\frac{x-1}{1-wx})} \qquad \Rightarrow \qquad \boxed{\mu_t = \frac{w\lambda t}{1-w}}$$

$$\boxed{\sigma_t^2 = \frac{w(1+w)\lambda t}{(1-w)^2}} \quad .$$

In this case, $\sigma^2 > \mu$ which is different from the pure Poisson process (where $\sigma^2 = \mu$).

If the conditions point to a Poisson distribution but the empirical data indicate that $\sigma^2 > \mu$, then this case may involve a mixed Poisson distribution.

Mixed Poisson Distribution
With the mixed Poisson distribution, the rate λ is itself distributed with

$\varphi(\lambda)$ dλ: (generalized) probability density function of λ.

Then

$$\boxed{P_u = \int_0^\infty \frac{\lambda^u}{u!} e^{-\lambda} \varphi(\lambda) \, d\lambda} \qquad , \qquad u \in \mathbb{N}_0 \quad . \qquad (19.9)$$

MIXED POISSON DISTRIBUTION

The generating function is

$$G(x)_{gem.P.} = \int_0^\infty e^{\lambda(x-1)} G(\lambda) \, d\lambda \quad .$$

$G(\lambda)$ is the generating function of the distribution of λ.

$$\mu = \int_0^\infty \lambda G(\lambda) \, d\lambda$$

$$\sigma^2 = \int \lambda G^2(\lambda) \, d\lambda < \mu, \text{ since } G^2 < G \text{ for } 0 < G < 1.$$

§20 GENERAL REMARKS ON CHANCE

A stroke of chance or coincidence is often associated with superstition or a flight into irrationality. Scientifically, "chance" is explained through the term probability. CHEVALIER DE MERE provided the impulse to deeper mathematical considerations when he asked PASCAL in a letter to make some statements about the prospects of winning in a prematurely ended card game. (RENYI (1969)).

A new situation arises in inventory: the earlier cost–profit decision criteria for the choice of the best course of action are now dependent on chance. In retrospect, it is then possible that the decision of an idiot would prove to be the best while that of an expert would be the worst. This, however, is an exception. In the long run, a valid decision rule will always prove to be better (according to the law of large numbers). "Only the competent have luck in the long run." Here lies the essential justification of OR in situations under risk.

For a process subject to chance, the extent of a desired result is typically determined by a specific course of action, i.e., a selected action, which controls the development of the process, is also subject to chance. Its probability distribution is known.

The choice of an action may be traced back to the corresponding probability distribution of the result. Hence, a criterion is needed for the choice of a distribution. Originally, one used the EXPECTED VALUE CRITERION.

Let

P_a : the distribution corresponding to action a of result x

Then, by the expected value criterion

$$a_1 \text{ is better than } a_2, \text{ if } E_{P_{a_1}}\{x\} > E_{P_{a_2}}\{x\}.$$

An objection against this criterion is the so–called PETERSBURG PARADOX:

One tosses a coin a number of times until a "head" appears. When this happens at the nth toss, one wins $x = 2^n$ from the bank. The distribution φ of winning with the action $a_1 =$ "play the Game" with prob(Tail) = prob(Head) = 0.5 has the infinite expected value

$$E_{P_{a_1}}\{x\} = \sum_{n=1}^{\infty} (\tfrac{1}{2})^n \cdot 2^n = 1 + 1 + \dots .$$

The action $a_2 =$ "do not play" does not give any return. According to the expected value criterion, a player must also be prepared to pay a large entry fee to the bank before he can play. In reality, no one is inclined to pay such a large amount.

The above paradox can be solved with the help of the criterion of expected utility which was introduced by and named after DANIEL BERNOULLI (1738). One measures the utility of money instead of the monetary payoff.

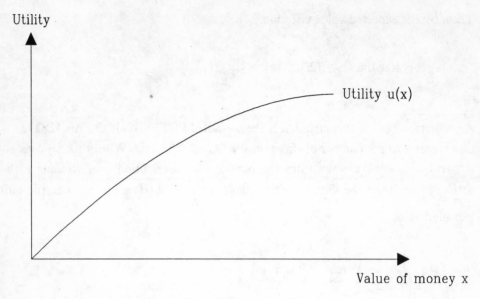

Figure 20.1: Utility function u

u(x) :   Utility of the result x

A utility function is rarely linear. In general, it is upper bounded. As long as it remains unbounded, new paradoxes can be constructed. In general economic terms, the utility function is assumed to be concave.

The expected utility is

$$E_{P_a} \{u(x)\} = \int u(x) dP_a(x) ,$$

and according to the utility criterion

$$a_1 \text{ is better than } a_2, \text{ if } E_{P_{a_1}} \{u(x)\} > E_{P_{a_2}} \{u(x)\} .$$

How does one know that the decisions based on the BERNOUILLI PRINCIPLE are good? One first tries to clearly derive, as much as possible, some enlightening consequences from this principle. As one finds more plausible consequences, the more plausible the Bernoulli principle becomes.

JOHN V. NEUMANN and OSKAR MORGENSTERN established an axiomatic system for rational behavior which implies the Bernoulli principle, the so–called "Utility Axioms". These axioms are plausible in themselves, although there are some doubts (ALLAIS).

A detailed discussion of decision theory under risk and uncertainty is found in CH. SCHNEEWEISS (1967) and DE GROOT (1970).

§21   INTEREST, CONTINUOUS INTEREST, PRESENT VALUE

Interest
Why is there interest? Obviously, a unit of money has more value today than one year later, even if the inflation rate is zero. The reason lies in the fact that the use of money brings with it a return, which can be paid as interest i. Discounting is used when interest goes back in time.

$$
\begin{array}{lll}
\text{Now} & & \text{in a Year} \\
1 & \xrightarrow{\quad \text{Interest} \quad} & 1 + i \\
\rho := \dfrac{1}{1+i} & \xleftarrow{\quad \text{Discount} \quad} & 1
\end{array}
$$

i   Interest;

1 + i: Interest Factor;

1/(1+i):   Discount Factor ρ; used to compute the present value of a future return.

Continuous Interest Payment

The annual interest rate is usually i. For semi–annual payments, capital grows by a factor of $(1+i/2)^2$, by n payments per year $(1+i/n)^n$. Taking limits, one obtains the continuous growth rate of capital

Continuous interest payment: $\lim\limits_{n\to\infty} (1 + \frac{i}{n})^n = e^i$.

Since

$$e^i = 1 + i + \frac{i^2}{2} + \dots > 1 + i$$

continuous compound interest is larger than discrete simple interest. An annual interest of i corresponds to an interest intensity r < i.

Interest Intensity r: $\boxed{1 + i = e^r}$ (21.1)

$$r = \ln(1 + i) = i - \frac{i^2}{2} + \frac{i^3}{3} \bar{} \dots$$

The discount factor ρ resulting from (21.1) is

$$\boxed{\rho = \frac{1}{1 + i} = e^{-r}} \quad .$$ (21.2)

Present Value

We now consider a flow of future payments which occur at equidistant time periods (year–end) t = 0,1,2,3,... For the decision problem it is necessary to evaluate the flow of payments relative to a specific time period. One usually chooses the last period or (often) the present time period. In the latter case, the so–called present value (instead of the final value) is computed. The present value is preferable for decisions in the current period.

Let

z_t :   Flow of payment, $t = 0,1,...T$

G :   Present value. It is defined as

$$G_\rho = \sum_{t=0}^{T} \frac{z_t}{(1 + i)^t} = \sum_{t=0}^{T} z_t \rho^t$$

Aside from the volume of money, the average payment

C :   Average value

is an important indicator of payment flow. C is the average payment per unit time

$$C = \frac{1}{T+1} \sum_{t=0}^{T} z_t .$$

There is a relationship between C and G. Thus, for large T and ρ very near 1, we have

$$z_0 + \rho z_1 + ... + \rho^T z_t \approx C(1 + \rho + ... + + \rho^T).$$

$$C \approx \frac{G}{1 + \rho + ... + \rho^T} = \frac{1 - \rho}{1 - \rho^{T+1}} G_\rho$$

For stationary models with an infinite planning horizon, i.e. all the z_t's are identical, the flow of payment is infinitely long. Then

$$\boxed{C_{(T=\infty)} = \lim_{\rho \to 1} (1 - \rho) G_\rho} .$$

(21.3)

§22   INVENTORY WITH POISSON DEMAND AND IMMEDIATE DELIVERY

One of the simplest stochastic inventory models is inventory with Poisson demand and immediate delivery. This model is interesting because it is handled with methods which are different from the previous models. Since a Poisson process is assumed for demand, the time since the occurrence of the last demand does not play a role.

Inventory is considered as a business which yields profit. The present value of future profits is dependent on the starting inventory y: G = G(y). We formulate G(y) recursive in time in which we divide the future into two parts: a small time period Δt lying immediately ahead and the rest of the periods. Because of the Poisson demand, it is not necessary to use t as an explicit argument of G. For all y we have

$$G(y) = \underbrace{-hy\Delta t}_{1)} + \underbrace{(1 - \lambda\Delta t)}_{2)}\underbrace{G(y)e^{-r\Delta t}}_{3)} +$$

$$+ \underbrace{\lambda\Delta t}_{4)}\underbrace{[b}_{5)} + Max\,\{\underbrace{-k - a(D - y + 1)}_{6)} + \underbrace{G(D)e^{-r\Delta t}}_{7)} \mid \underbrace{G(y-1)e^{-r\Delta t}}_{8)}\}] \qquad (22.1)$$

1) Inventory costs during Δt
2) Probabity that no demand occurred in Δt
3) Present value after Δt
4) Probability that demand occurred in Δt
5) Sales Revenue
6) Order Costs
7) Present value after Δt, if an order occurred in Δt
8) Present value after Δt, if no order occurred in Δt

Note: With a Poisson demand, only one unit is demanded per event.

The recursion (22.1) formulates the "Principle of Optimality" of dynamic programming (BELLMAN's Principle of Optimality; BELLMAN (1957), BECKMANN (1968)).

This equation is explained as follows:

The present inventory is y. The present value G(y) is the value of all future costs and revenues based on the present. It is written on the right side of (22.1) but is now separated into the time span Δt and the rest of the periods. At first, inventory costs accumulate in a small time span Δt (Term 1). At the end of Δt time periods we add the profits from the rest of the periods to the accruing inventory costs. This value depends on whether or not an order occurs after Δt and, in case demand occurs, whether or not one orders.

Case 1:	No demand occurs: Terms 2) and 3)
Case 2:	Demand occurs with probability $\lambda \Delta t$
	Case 2a: Stock up to D: Terms 6) and 7)
	Case 2b: do not order: Term 8)

In any case the future costs and profits of the remaining terms must also be discounted into the present; hence, the factor $e^{-r\Delta t}$. The inventory cost hy is not discounted within the time period Δt (one may interpret the inventory cost so that the discounting, i.e., the interest cost, is already contained in h).

The solution of the functional equation (22.1) determines a corresponding optimal decision for each y, i.e. the action which yields the maximum on the right hand side of (22.1). Since the functional equation is solved for all valid y, one obtains for each y an optimal course of action and hence, as a whole, a _decision rule_ or policy.

In the given case the decision rule is already prestructured. D is no longer the lot size, but the amount which must be kept in stock. On the other hand, an order should be placed only when the inventory sinks to zero. If one places an order at $y_0 > 0$, then one would hold a constant "floor stock" y_0 which would never be used. Because of this, D is again the lot size and from (22.1)

$$G(y) = -hy\Delta t + (1 - \lambda\Delta t)\, G(y)e^{-r\Delta t} + \lambda\Delta t\,(b + G(y-1)e^{-r\Delta t}),\ y > 1, \qquad (22.2)$$

$$G(1) = -h\Delta t + (1 - \lambda\Delta t)\, G(1)e^{-r\Delta t} + \lambda\Delta t\,(b - k - aD + G(D)e^{-r\Delta t}). \qquad (22.3)$$

(22.3) is the boundary condition to the differential equation (22.2).

If one approximates $e^{-r\Delta t}$ by $1 - r\Delta t$, one obtains

$$G(y) = -hy\Delta t + G(y) - G(y)(\lambda + r)\Delta t + \lambda\Delta tb + \lambda\Delta tG(y - 1) + o(\Delta t)^2$$

$$G(1) = -h\Delta t + G(1) - G(1)(\lambda + r)\Delta t + \lambda\Delta t(b - k - aD) + G(D)\lambda\Delta t + o(\Delta t)^2 .$$

With $\rho := \dfrac{\lambda}{\lambda + r}$, and omitting the term $o(\Delta t)^2$ from the above equation, we get

$$G(y) = -\frac{\rho}{\lambda} hy + \rho b + \rho G(y - 1) \ , \tag{22.4}$$

$$G(1) = -\frac{\rho}{\lambda} h + \rho b + \rho(-k - aD + G(D)) \ . \tag{22.5}$$

Equation (22.4) is a differential equation of order 1 with the boundary condition (22.5). The solution of this differential equation is obtained through successive substitution

$$G(D) \ = \rho b - \frac{\rho hD}{h} + \rho(\rho b - \frac{\rho h(D - 1)}{\lambda} + \rho(\rho b - \frac{\rho h(D - 2)}{\lambda} +$$

$$+ \ldots + \rho(\rho b - \frac{\rho}{\lambda}h \cdot 2 + \rho G(1))\ldots))$$

$$= \frac{1}{1 - \rho^D}\left\{\rho b \frac{1 - \rho^D}{1 - \rho} - \frac{h\rho}{\lambda} \sum_{i=0}^{D-1} (D - i)\rho^i - \rho^{D-1}(k + aD)\right\}$$

$$= \frac{\rho b}{1 - \rho} - \frac{\rho}{1 - \rho^D}\left\{\frac{h}{\lambda} \sum_{i=0}^{D-1} (D - i)\rho^i + \rho^{D-1}(k + aD)\right\}. \tag{22.6}$$

To interpret this formula, let

$\frac{1}{\lambda}$: Average interval between two demands

$\frac{r}{\lambda}$: Applicable interest rate for the interval $\frac{1}{\lambda}$

$\rho = \dfrac{1}{1 + \frac{r}{\lambda}}$: Discount factor for the time interval $\frac{1}{\lambda}$

ρ^D : Discount factor for one cycle

$\dfrac{b}{1 - \rho^D}$: Present value of total profits

$\dfrac{1}{1 - \rho^D}\{\ \}$: Present value of all cycle costs

$\dfrac{h}{\lambda} \displaystyle\sum_{i=0}^{D-1} (D{-}i)\rho^i$: Average inventory cost of a cycle (discounted within the cycle)

Assume we are in decision period zero when there is no inventory in stock. The equation

$$G(0) = -k - aD + G(D)$$

applies.

We substitute G(D) from (22.6) and obtain

$$G(0) = \frac{-k - aD - \frac{h\rho}{\lambda} \sum (D - i)\rho^i}{1 - \rho^D} + \rho\,\frac{b}{1 - \rho} \ . \tag{22.7}$$

The numerator Z of the first fraction on the right hand side of (22.7) represents the cost per cycle. $Z/(1 - \rho^D) = Z(1 + \rho^D + (\rho^D)^2 + ...)$ is the present value of all cycle costs. The term $\rho b/(1 - \rho)$ is the present value of total profits. (Note: No demand is lost). It is independent of D.

Only the costs are dependent on D. These indicate that one could have easily formulated the problem as a cost minimization problem instead of a profit maximization problem. We now want to see from (22.7) exactly how the cost minimization problem proceeds.

The summation $\sum\limits_{i=0}^{D-1} (D-i)\rho^i$ can be converted to

$$\sum_{i=0}^{D-1} (D-i)\rho^i = \sum_{j=0}^{D-1} \rho^j + \sum_{j=0}^{D-2} \rho^j + \ldots + \sum_{j=0}^{0} \rho^j =$$

$$= \frac{1-\rho^D}{1-\rho} + \frac{1-\rho^{D-1}}{1-\rho} + \ldots + \frac{1-\rho}{1-\rho} =$$

$$= \frac{1}{1-\rho}\left(D - \frac{\rho}{1-\rho}(1-\rho^D)\right).$$

From this, (20.7) becomes

$$G(0) = \rho\underbrace{\frac{b}{1-\rho}}_{\text{const.}} - \frac{k+aD}{1-\rho^D} - \frac{h\rho}{\lambda}\frac{D}{(1-\rho)(1-\rho^D)} + \underbrace{\frac{h\rho^2}{\lambda}\frac{1}{(1-\rho)^2}}_{\text{const.}} . \qquad (22.8)$$

The last term in this equation is a constant which is exactly the present value of the profit $\rho b/(1-\rho)$ such that (22.8) takes the form

$$G(0) = \text{Constant} - C_1$$

The term C_1, which is dependent on D, contains all negative terms of the profit function $G(0)$. It represents, therefore, all costs.

The resulting cost minimization problem , with k + aD =: K, is

$$C_1 = \frac{K\lambda(1 - \rho) + h\rho D}{\lambda(1 - \rho)(1 - \rho^D)} \rightarrow \underset{D}{\text{Min}} \quad ,$$

or after simplification

$$C = \frac{K\lambda(1 - \rho) + h\rho D}{1 - \rho^D} \rightarrow \underset{D}{\text{Min}} \quad . \tag{22.9}$$

C is convex. Hence, the minimum is determined by dC/dD = 0. We substitute

$$\rho := e^{-r}$$

and obtain

$$\frac{dC}{dD} \overset{!}{=} 0 \quad \Leftrightarrow \quad [a\lambda(1 - \rho) + h\rho](1 - \rho^D) - r\rho^D[(k + aD)\lambda(1 - \rho) + h\rho D] \overset{!}{=} 0$$

$$a\lambda(1 - \rho)[1 - \rho^D - rD\rho^D] + h\rho[1 - \rho^D - rD\rho^D] = r\rho^D k\lambda(1 - \rho)$$

$$[a\lambda(1 - \rho) + h\rho][1 - \rho^D - rD\rho^D] = r\rho^D k\lambda(1 - \rho)$$

$$\rho^{-D} - 1 - rD = \frac{rk\lambda}{a\lambda + h\dfrac{\rho}{1 - \rho}}$$

$$e^{rD} - 1 - rD = \frac{rk\lambda}{a\lambda + h\dfrac{1}{e^r - 1}} \quad . \tag{22.10}$$

We expand e^{rD} into a Taylor Series

$$\frac{r^2 D^2}{2} + \frac{r^3 D^3}{3!} + \ldots = \frac{rk\lambda}{a\lambda + h\dfrac{1}{e^r - 1}} \quad .$$

For $r << 1$, $e^r - 1 \approx r$ and one obtains the estimate

$$\frac{r^2 D^2}{2} \approx \frac{r^2 k\lambda}{a\lambda r + h}$$

$$\boxed{D \approx \sqrt{\frac{2\lambda k}{a\lambda r + h}}} \quad . \tag{22.11}$$

This result shows that the interest–created effect can be interpreted as an increase in the inventory holding cost rate from h to $a\lambda r + h$. The higher the interest rate is, the smaller is the optimal lot size.

In limits, as $\rho \to 1$

$$D = \sqrt{\frac{2\lambda k}{h}} \quad . \tag{22.12}$$

The result in the undiscounted case may also be directly derived from (22.8). From (21.3) we know that

$$C_{(T = \infty)} = \lim_{\rho \to 1} (1 - \rho) G_\rho.$$

Hence

$$\lim_{\rho \to 1} (1 - \rho) G_\rho(0) = \lim_{\rho \to 1} (1 - \rho) \cdot \left[\frac{\rho b}{1 - \rho} - \frac{k + aD + \dfrac{h}{\lambda} \displaystyle\sum_{i=0}^{D-1} (D-i)\rho^i}{1 - \rho^D} \right] \quad .$$

$$(1 - \rho) G_\rho(0) = \rho b - \frac{k + aD + \dfrac{h}{\lambda} \displaystyle\sum_{i=0}^{D-1} (D - i)\rho^i}{1 + \rho + \rho^2 + \ldots + \rho^{D-1}}$$

$$\lim_{\rho \to 1} (1 - \rho) G_\rho(0) = b - \frac{k}{D} - a - \frac{h}{\lambda} \cdot \frac{(D + 1)}{2}$$

$$= \text{Constant} - C_{1, \rho = 1} \quad ,$$

and we obtain, as in the deterministic model, the equivalent cost minimization problem

$$\underset{D}{\text{Min}} \ C_{1,\rho=1} = \underset{D}{\text{Min}} \ \{ \frac{k}{D} + \frac{h(D+1)}{2\lambda} \} \ ,$$

whereby (22.12) also follows.

§23   POISSON DEMAND, NO DISCOUNTING

In the previous sections, we discussed the case of no discounting by the "backdoor" method of the limiting value $\rho \to 1$. Now we want to formulate the corresponding model with the help of the Principle of Optimality.

The case of no discounting contains conceptual difficulties since some of the present values of revenues and costs become infinitely large. In this case the minimization of the growth rate of cost represents a suitable objective function. The resulting model is discussed in this section.

Under the present assumption that no shortages are allowed, it is reasonable to ignore the sales profit. Since discounting is ignored, shifting the time period for realizing profits does not have any effect. The only important thing is the total profit. Since the total revenues from the inventory control policy are not influenced, we choose a cost approach.

Let

Θ : Planning horizon

$l_\Theta(y)$: Loss function with inventory y and planning horizon Θ

$l(y)$: $\underset{\Theta \to \infty}{\lim} l_\Theta(y)$, if it exists

Since, in the undiscounted case, the costs are proportional to time t in the long run, $l_\Theta(y)$ will increase asymptotically linear for very large Θ.

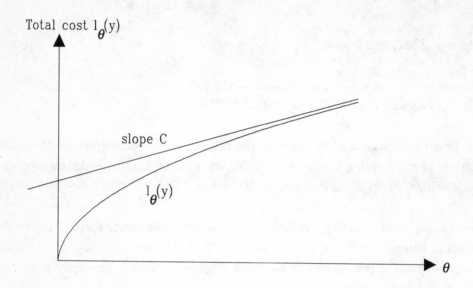

Figure 23.1: Asymptotically linear total cost

Therefore, in the stationary case,

C :   Cost growth rate per unit time

is a constant value. For $\Theta \to \infty$,

$$l_\Theta(y) = C\Delta t + l_{\Theta - \Delta t}(y) \; . \tag{23.1}$$

From the recursive method (Note: the planning horizon is shortened with increasing calendar time)

$$l_\Theta(y) = hy\Delta t + (1 - \lambda\Delta t)l_{\Theta-\Delta t}(y) + \lambda\Delta t l_{\Theta-\Delta t}(y - 1) , \quad y > 1$$

$$l_\Theta(1) = h\Delta t + (1 - \lambda\Delta t)l_{\Theta-\Delta t}(1) + \lambda\Delta t[k + aD + l_{\Theta-\Delta t}(D)] , \tag{23.2}$$

becomes

$$C\Delta t + l_{\Theta-\Delta t}(y) = hy\Delta t + (1 - \lambda\Delta t)l_{\Theta-\Delta t}(y) + \lambda\Delta t l_{\Theta-\Delta t}(y - 1), \, y > 1$$

$$C\Delta t + l_{\Theta-\Delta t}(1) = h\Delta t + (1 - \lambda\Delta t)l_{\Theta-\Delta t}(1) + \lambda\Delta t[k + aD + l_{\Theta-\Delta t}(D)]. \tag{23.3}$$

It is again assumed here that there will be an order only at $y = 0$. We must show that this naive method is successful, i.e., whether reasonable results for l and D can be derived from it.

We represent (23.3) in the form

$$C\Delta t + \lambda\Delta t l_{\Theta-\Delta t}(1) = h\Delta t + \lambda\Delta t(k + aD) + \lambda\Delta t l_{\Theta-\Delta t}(D)$$

$$C\Delta t + \lambda\Delta t l_{\Theta-\Delta t}(2) = 2h\Delta t + \lambda\Delta t l_{\Theta-\Delta t}(1)$$

$$\vdots \tag{23.4}$$

$$C\Delta t + \lambda\Delta t l_{\Theta-\Delta t}(D) = Dh\Delta t + \lambda\Delta t l_{\Theta-\Delta t}(D - 1) .$$

By summation of these equations, the first terms drop out. What remains is

$$DC = h \sum_{i=1}^{D} i + \lambda(k + aD) ,$$

hence

$$C = \frac{hD(D + 1)}{2} + \frac{\lambda k}{D} + \lambda a \; .$$

(23.5)

This is the stationary cost rate (cost of cycle per unit time). It is necessary to minimize:

$$\frac{hD(D + 1)}{2} + \frac{\lambda k}{D} + \lambda a \to \underset{D}{\mathrm{Min}} \; .$$

We already have the same objective function in the deterministic model. Hence, the WILSON formula also follows with Poisson demand without discounting

$$\boxed{D^* = \sqrt{\frac{2\lambda k}{h}}} \; .$$

(23.6)

How does one interpret C in a stochastic sense?

The different stocks $y = 1, 2, ..., D$ are the possible states of the system. The probability π_y, with which the system finds itself in state y, is the same for all states since demand rate λ is constant

$$\pi_y = \frac{1}{D}, \qquad y = 1, 2, ..., D.$$

We now write (23.5) in the form

$$C = h \sum_{y=1}^{D} y \, \frac{1}{D} + \frac{1}{D} \lambda (k + aD) \; .$$

(23.7)

$$\underbrace{\qquad}_{1)} \;\; \overset{\uparrow\;\uparrow}{2)\,3)}$$

1) Inventory is y with probability 1/D
2) Probability of finding the system in state y = 1
3) y = 0 with rate λ and the cost k + aD is incurred

so that (23.7) can be interpreted as the expected value of the costs of a cycle

$$C = \sum_{y=1}^{D} c_y \pi_y$$

c_y : Average cost of a cycle per unit time in state y.

The state probabilities π_y (in the present case $\pi_y = \frac{1}{D}$) are dependent on the lot size D. The method, which determines D by the minimization of (23.7), is called the METHOD OF STATE PROBABILITIES (more in section §31).

It is interesting to note that, in the present case, both methods lead to the same objective function but in different forms.

§24 RECURRENT PROCESS

Now let $\lambda = \lambda(t)$ be dependent on the time which has elapsed since the last event. This situation, for example, can occur in a newsstand which is located at a bus stop. The customers are mainly the bus passengers. As a rule, the bus is not punctual. The interval between two arrivals is then stochastic. Under the above assumption, the probability of arrival is dependent on the time since the last arrival. The longer it takes for one to wait for the bus, the higher is the probability that it will arrive at any moment.

The state space is now two—dimensional: the inventory y and the time t since the last demand. Let

$l_\Theta(y,t)$: Cost function with starting inventory y and planning horizon Θ, whereby time t has elapsed since the last demand.

We assume that the cost function increases linearly with time for a fixed y for $\Theta \to \infty$ and with a rate C. Then the method "total cost tomorrow = total cost today + C" which is analogous to (23.1) leads to

$$l_\Theta(y, t + \Delta t) = l_{\Theta - \Delta t}(y, t) + C\Delta t \ . \tag{24.1}$$

Corresponding to (23.2), the recursive method is

$$l_\Theta(y, t + \Delta t) = hy\Delta t + [1 - \lambda(t)\Delta t] l_{\Theta - \Delta t}(y, t + \Delta t) +$$

$$+ \lambda(t)\Delta t \ \underset{\underset{x>0}{\uparrow}}{Min}\{l_{\Theta - \Delta t}(y - 1, 0), \ \underset{y>1}{Min}\{k + ax + l_{\Theta - \Delta t}(x, 0)\} \ , \tag{24.2}$$

since demand has now occurred

$$l_\Theta(1, t + \Delta t) = h\Delta t + [1 - \lambda(t)\Delta t] l_{\Theta - \Delta t}(1, t + \Delta t) +$$

$$+ \lambda(t)\Delta t \ \underset{x>0}{Min} \{k + ax + l_{\Theta - \Delta t}(x, 0)\} \ . \tag{24.3}$$

Since cost increases are stationary ($\Theta \to \infty$) we need not consider the planning horizon Θ explicitly and, therefore, may drop the index Θ or $\Theta - \Delta t$.

The functional equations (24.2), (24.3) are discrete in y and continuous in t. The following computation shows that it can be transformed for $\Delta t \to 0$ such that the cost function l is dependent only on y at time period t = 0. The process is interesting only at the transition periods (these are the renewal points for the function $\lambda(t)$ where $\lambda(t)$ is reset to the starting value $\lambda(0)$). Such a process is called a <u>recurrent process</u>.

We again have the same assumption: Order quantity $x(y) = \begin{cases} 0 & \text{for } y > 0, \\ D & \text{for } y = 0. \end{cases}$

Then from (24.2) and using (24.1):

$$\frac{-1(y,\ t\ +\ \Delta t)\ +\ 1(y,t)}{\Delta t} + \lambda(t)1(y,t + \Delta t) = hy - C + \lambda(t)1(y-1,0)$$

and for $\Delta t \rightarrow 0$, we have the linear differential equation

$$-\frac{\partial 1(y,t)}{\partial t} + \lambda(t)1(y,t) = hy - C + \lambda(t)1(y-1,0) \ . \tag{24.4}$$

Intermediate computation: we solve (24.2) by integration. Again (24.4):

$$-\dot{1} + \lambda(t)1 = hy - C + \lambda(t)1(y-1,0) \ .$$

The integration becomes easy if the left hand side of the integral is a product $1 \cdot f$. To achieve this, we multiply (24.4) with $f(t)$, the so–called integrating factor

$$f(t) = e^{\displaystyle -\int_0^t \lambda(x)dx} \ .$$

The left hand side of the above equation becomes

$$-\dot{1}f + 1\lambda f, $$

and because of the special form of f, it is identical to the derivative of $1f$

$$-\dot{1}f + 1\lambda f = -\frac{\partial}{\partial t}(1f) \ .$$

Hence (24.4) becomes

$$-\frac{\partial}{\partial t}(1f) = [hy - C + \lambda(t)1(y-1,0)]f \ ,$$

and only the right hand side remains to be integrated. One obtains by using partial integration

$$\int_0^\infty f(t)\,dt = \underbrace{\int_0^\infty 1 \cdot e^{-\int_0^t \lambda(x)\,dx}}_{1)} = tf(t)\,\big|_0^\infty + \underbrace{\int_0^\infty t\lambda(t)e^{-\int_0^t \lambda(x)\,dx}}_{2)}\,dt\ .$$

$$\underbrace{\phantom{\int_0^\infty 1 \cdot e^{-\int_0^t \lambda(x)\,dx} = tf(t)\,\big|_0^\infty + \int_0^\infty t\lambda(t)e^{-\int_0^t \lambda(x)\,dx}\,dt}}_{3)\ =:\ \alpha}$$

1) Probability that no event occurs until time t

2) Probability that an event occurs in time t

3) Expected value of the time interval until the next event

The integration of the complete equation results in

$$-1f\big|_0^\infty = (hy - C)\alpha + 1(y - 1,0)\int_0^\infty \lambda(t)e^{-\int_0^t \lambda(x)\,dx}$$

$$-1f\big|_0^\infty = (hy - C)\alpha + 1(y - 1,0)(-1)e^{-\int_0^t \lambda(x)\,dx}\ \big|_0^\infty\ .$$

We let $1(y,\infty)\cdot 0 = 0$ and obtain as a solution from (24.4)

$$1(y,0) = (hy - C)\alpha - 1(y - 1,0)\ .$$

With this, the time t is eliminated and we get the intermediate result

$$1(y) = (hy - C)\alpha - 1(y - 1)\ . \tag{24.5}$$

If one expands the recursion (24.5), one obtains

$$1(y) = \alpha(hy - C) + 1(y - 1) =$$

$$= \alpha(hy - C) + (h(y - 1) - C) + [\ldots + \alpha(2h - C + 1(1))]\ \ldots]]\ ,$$

and therefore

$$l(y) = \alpha \cdot \sum_{i=2}^{y} (hi - C) + l(1) \ , \qquad 1 < y \le D \tag{24.6}$$

$$l(1) = \alpha(h - C) + k + aD + l(D) \ .$$

(24.6) is a system of D equations in D+1 unknowns l(1), l(2), ..., l(D), C. The lot size D is considered as given and later determined by minimization. From the chosen optimization criterion "minimize the stationary cost increase per unit time", it follows that the optimal D is dependent only on the relative values of l. Hence, since one of the l(y) can be arbitrarily chosen, we set

$$l(1) := \alpha(h - C)$$

and obtain

$$l(y) = \alpha \sum_{i=1}^{y} (hi - C)$$

or

$$\boxed{\begin{aligned} l(y) &= \frac{\alpha hy(y + 1)}{\lambda} - \alpha yC \quad , \ 1 < y \le D \ . \\[2mm] l(1) &= \alpha(h - C) \end{aligned}} \tag{24.7}$$

For a Poisson Demand

$$\alpha = \int_{0}^{\infty} t\lambda e^{-\lambda t} dt = \frac{1}{\lambda} \ ,$$

and (24.7) becomes

$$l(y) = \frac{hy(y + 1)}{2\lambda} - \frac{yC}{\lambda} , \ 1 < y \le D \ , \tag{24.8}$$

$$l(1) = \frac{1}{\lambda}(h - C).$$

This specification, however, is not essential.

$l(y)$ is the value function of dynamic programming. By summation of the equation (24.6) one eliminates the $l(y)$ and, in this case (compare (23.5)), obtains also the cost function

$$C = \frac{h(D + 1)}{2} + \frac{\lambda k}{D} + \lambda a \ , \tag{24.9}$$

from which the optimal lot size results

$$\boxed{D^* = \sqrt{\frac{2\lambda k}{h}}} \tag{24.10}$$

§25   PROOF OF OPTIMALITY

The above results were derived under the assumption that the optimal ordering rule of the form "order D, in case y=0" has a given fixed structure. Optimization was carried out only under this type of an ordering rule. Assumptions and results seem to be feasible. Still missing, however, is the proof which shows that the optimal ordering rule has such a given structure.

An optimal value function which does not assume any specific ordering rule obeys the functional equation

$$l(y) + C\Delta t = hy\Delta t + [1 - \lambda\Delta t]l(y) + \lambda\Delta t \underset{x}{\mathrm{Min}} \{l(y - 1), \underset{x}{\mathrm{Min}} \{k + ax + l(x)\}\}, \ y > 1,$$

$$\tag{25.1}$$

$$l(1) + C\Delta t = h\Delta t + [1 - \lambda\Delta t]l(1) + \lambda\Delta t \underset{x}{\mathrm{Min}} \{k + ax + l(x)\} \ .$$

We now show that our results (24.8), (24.9), (24.10) satisfy these
functional equations. We substitute l,C and D from (24.8), (24.9) and (24.10) into
(25.1).

<u>l(x) substituted:</u> The minimization Min { } gives

$$\underset{x}{\text{Min}}\{k + ax + \frac{h(x + 1)x}{2\lambda} - \frac{xC}{\lambda}\} \qquad (\text{convex!})$$

$$\frac{d}{dx} \overset{!}{=} 0: \quad a + \frac{h}{\lambda}x + \frac{h}{2\lambda} - \frac{C}{\lambda} = 0$$

$$x = \frac{C}{h} - \frac{1}{2} - a\frac{\lambda}{h} \quad.$$

C substituted from (24.9):

$$x = \frac{D + 1}{2} + \frac{\lambda a}{h} + \frac{\lambda k}{hD} - \frac{1}{2} - \frac{\lambda}{h}a$$

$$x = \frac{1}{2}(D + \frac{2k\lambda}{hD}) \quad. \tag{25.2}$$

D substituted from (24.10):

$$x = \frac{1}{2}(D + \frac{D^2}{D}) = D \quad.$$

Hence

$$\underset{x}{\text{Min}} \{k + ax + l(x)\} = 0$$

and

$$\underset{x}{\text{Min}} \{l(y-1),0\} = l(y-1) \ , \qquad y > 1 \ ,$$

since

$$l(y) = \frac{y}{\lambda} [\underbrace{\frac{h(y + 1)}{2} - C}_{<0}] \ , \qquad y > 1 \ .$$

Hence, the ordering rule "order D, if y = 0" satisfies the principle of optimality (25.1). It remains to be shown that it is the only solution to the functional equation (25.1).

Let $x = D' < \sqrt{\frac{2\lambda k}{h}}$ be another order quantity. Substituting D'in (25.2)

$$D' = \frac{1}{2}(D' + \frac{D^2}{D'}) , \qquad\qquad D' < D$$

leads to the contradiction

$$D' = \frac{D^2}{D'} > D . \qquad\qquad\qquad (25.3)$$

The assumption $x = D' > D$ also leads to a contradiction.

Remark 1: $D^* = \sqrt{\frac{2\lambda k}{h}}$ is the only real–valued solution to (25.1). However, since y and D are limited to integral values, two neighboring order quantities D_1 and D_2 ($D_2 = D_1 + 1$) are optimal.

Remark 2:   It can be shown that the computational steps

   1) Choose D'
   2) Compute $l(y)|D'$, and C(D') from (24.8), (24.9) and (24.10)
   3) Compute $x = x(D')$ from (25.2)

   lead to an improvement "x(D') is better than D'", i.e., C(x(D')) < C(D') as long as $D' \neq D^*$. The optimal solution D^* is finally obtained after a finite number of improvements (only requirement: h, k and a are all non–negative).

This method of dynamic programming is called POLICY ITERATION. A very detailed explanation of these methods are found in BECKMANN (1968) and HOWARD (1965).

CHAPTER 3:
STOCHASTIC SINGLE PERIOD MODELS

§26   THE NEWSBOY PROBLEM

Model with Proportional Shortage Cost

The inventory models we have discussed to this point have been characterized by continuous stock monitoring. An order may be placed at any point in time. In contrast to these are the periodic models. Stock inspection and/or orders are possible only at discrete points in time, i.e., at the beginning of a period. If nothing is explicitly specified, all periods are taken to be of the same length.

The simplest periodic model is the single period model. The decision problem reduces to only one period. Such inventory problems occur if the products cannot be sold after the period. Examples of these are fashion articles, travel offers, ticket sales for large presentations and daily newspapers. We formulate the last case as the basic model of the newsboy problem (or, in more recent terminology, the newspaper vending machine problem).

Early in the morning, the newsboy buys a stack of newspapers and tries to sell these during the course of the day. He can only return the unsold papers at a loss. If he carries only a small quantity of newspapers, he misses out on a profit. Demand is uncertain but its distribution is known. His decision problem: "How many newspapers do I buy to maximize my profit expectations?"

Let

x :   Stock of newspapers which the newsboy carries early in the morning

p_u :   Probability that u copies will be sold

$P(u)$:   Probability that the demand is actually smaller than u

f :   Expected value of u

h :   Loss per unsold copy because of surplus

g :   Loss per unsold copy because of shortage (loss of profit and customer dissatisfaction)

In a typical decision situation, g >> h.

The decision variable is the initial stock x. The problem can be simplified under the non–limiting assumption that all possible shortages occur only at the end of a period. It is, therefore, enough to consider the situation at the end of the period. The time distribution of profit during the period can be ignored. The value x is to be selected such that the expected utility of the situation at the end of the period is maximized:

$$\text{Max}_x \ E\{\text{Utility at the end of the period}\}.$$

The utility function has the following form:

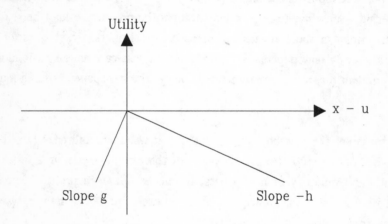

Figure 26.1: Utility function of the newsboy at the end of a period

The objective function is then

$$\text{Max}_x \ \{-h \sum_{u=0}^{x} (x-u)p_u - g \sum_{u=x+1}^{\infty} (u-x)p_u\}$$

or

$$\text{Min}_x \{ h \sum_{u=0}^{x} (x-u) p_u + g \sum_{u=x+1}^{\infty} (u-x) p_u \}. \qquad (26.1)$$

Since it is not convenient in practice to work with infinite summations, it is, therefore, useful to rewrite the objective function. We replace the summation by integration

$$\text{Min}_x \{ h \int_0^x (x-u) dP(u) + g \int_x^\infty (u-x) dP(u) \} =$$

$$\text{Min}_x \{ (h+g) \int_0^x (x-u) p_u du + g(\mu - x) \} .$$

Using partial integration

$$(h+g) \int_0^x (x-u) p_u du = \underbrace{(h+g)(x-u)P(u) \Big|_0^x}_{= 0, \text{ since } P(u) = 0} + (h+g) \int_0^x P(u) du$$

one obtains the objective funciton

$$\boxed{\text{Min}_x \{ (h+g) \int_0^x P(u) du + g(\mu - x) \}} \qquad . \qquad (26.2)$$

It is convex since $\dfrac{d^2}{dx^2} \int_0^x P(u) du = p_x \geq 0$.

From $\dfrac{d}{dx} \{ \} \overset{!}{=} 0$, it follows that

$$(h+g)P(x) - g = 0$$

$$\boxed{x = P^{-1}(\frac{g}{h+g})} \qquad . \qquad (26.3)$$

The solution is easy to determine. One need not know the distribution function P. The information about P near the point $\frac{g}{h+g}$ is enough. The simplest manner of determining x is the following:

Since $\qquad\qquad P(x) = \frac{g}{h+g}$

then $\qquad\qquad 1 - P(x) = \frac{h}{h+g}$.

Let g = 10h. Then $1 - P(x) = \frac{1}{11} = 9\%$ Hence, x must be chosen such that one carries fewer newspapers 9% of the time.

We now ask:

How large must the demand be so that it pays off to have a supply with initial inventory x > 0 ? An extreme solution x = 0 occurs if g/(h+g) reaches the critical value P(0). The business is profitable only if

$$\frac{g}{h+g} > P(0) .$$

In the above example, if demand does not occur at least 91% of the time, then one should give up the business.

Model with non–proportional shortage cost

To simplify the computation, the product to be stored in inventory is treated as a continuous variable (e.g., oil). The first applications of operations research and statistics were actually related to the provisioning of ships; for example, with fuel for a long journey. In these cases, it does not make too much sense to evaluate shortages with proportional costs. If three or five units of an important part are missing while on the high seas, then both cases are equally bad. Hence, we use the following approach:

$$\text{Min}_{x} \left\{ h \int_{0}^{x} (x-u)\,dP(u) + G \int_{x}^{\infty} dP(u) \right\} . \tag{26.4}$$

G : Constant arbitrary cost when a shortage occurs

The optimal lot size x is determined from the condition $\frac{d}{dx} \{ \ \} \stackrel{!}{=} 0$, i.e.

$$hP(x) - Gp_x = 0 .$$

Let the demand, for example, be exponentially distributed with the expected value $1/\lambda$: $P(x) = 1 - e^{-\lambda x}$. Then

$$h(1 - e^{-\lambda x}) - G\lambda e^{-\lambda x} = 0$$

$$h = (h + G\lambda)e^{-\lambda x}$$

$$x = \frac{1}{\lambda} \cdot \frac{\ln(h + G\lambda)}{\ln h} . \tag{26.5}$$

By a suitable choice of unit, $h > 1$ is always reached. Hence, (26.5) states that the stock quantity must always be greater than the expected consumption.

§27 EVALUATION OF $P(x) = \frac{g}{h + g}$

One of the most important demand distributions which occurs in practice is the Poisson distribution (see §19). If one considers the occurrence of demand during longer time periods, then the Poisson distribution is transformed into a normal distribution with the density

$$p(x)\,dx = \frac{1}{\sqrt{2\pi}\,\sigma}\, e^{-\frac{1}{2} \cdot \frac{(x-\mu)^2}{\sigma^2}}\, dx$$

μ: Expected value

σ^2:   Variance

The normal distribution can be approximated by the

LOGISTIC: $P(x) = \dfrac{1}{1 + e^{-mx}}$; $m \approx 1.6$.

$$P(x,\mu,\sigma) = \frac{1}{1 + e^{-\frac{m(x-\mu)}{\sigma}}} .$$ (27.1)

The value $m \approx 1.6$ is achieved as follows: The density of the standard normal distribution at $x = 0$ is $1/\sqrt{2\pi}$. The density of the standard logistic at $x = 0$ is

$$\frac{d}{dx} \frac{1}{1 + e^{-mx}} \Big|_{x=0} = \frac{me^{-mx}}{(1 + e^{-mx})^2} \Big|_{x=0} = \frac{m}{4} .$$

Since both densities should be equally large, it follows that

$$m = \frac{4}{\sqrt{2\pi}} \approx 1.6 .$$

For the newsboy problem, the condition for the optimal lot size when applying this approximation is

$$\frac{1}{1 + e^{-\frac{m(x-\mu)}{\sigma}}} = \frac{g}{h + g} = \frac{1}{1 + \frac{h}{g}}$$

and, hence

$$e^{\frac{m(x-\mu)}{\sigma}} = \frac{g}{h} ,$$

$$\boxed{x = \mu + \frac{\sigma}{m} \ln \frac{g}{h}}$$ (27.2)

The optimal lot size x is a linear function of μ and σ and an increasing function of g/h. One can see that

$$g\left\{\begin{matrix}>\\<\end{matrix}\right\}h \;\Rightarrow\; x\left\{\begin{matrix}>\\<\end{matrix}\right\}\mu \;. \tag{27.3}$$

We now investigate the costs. Let

l(x) :   Expected value of the costs for the single period model with optimal lot size x.

For a model with proportional shortage costs, the expected costs according to (26.2) are

$$l(x) = (h + g) \int_0^x P(u)\,du + g(\mu - x) \;, \tag{27.4}$$

and in the special case of logistically distributed demand

$$l(x) = (h + g) \int_{-\infty}^{x} \frac{1}{1 + e^{-\frac{m}{\sigma}(y-\mu)}}\,dy + g(\mu - x) =$$

$$= (h + g) \frac{\sigma}{m} \int_0^{x\frac{m}{\sigma}} \frac{e^{\frac{m}{\sigma}(y-\mu)}}{1 + e^{\frac{m}{\sigma}(y-\mu)}}\,dy + g(\mu - x) =$$

$$= (h + g) \frac{\sigma}{m} \ln\left[1 + e^{\frac{m}{\sigma}(x-\mu)}\right] + g(\mu - x) \;.$$

When applying the logistic, one assumes that a negative demand can be ignored.

If the expression (27.2) is substituted for the optimal x, one obtains

$$l(x) = (h + g) \frac{\sigma}{m} \ln[1 + e^{\ln \frac{g}{h}}] - g \frac{\sigma}{m} \ln \frac{g}{h} =$$

$$= \frac{\sigma}{m} [(h + g) \ln \frac{h + g}{h} - g \ln \frac{g}{h}] =$$

$$= \frac{\sigma}{m} [h \ln \frac{h + g}{h} + g \ln \frac{h + g}{g}]$$

and, finally,

$$l(x) = (h + g) \frac{\sigma}{m} [-\frac{h}{h + g} \ln \frac{h}{h + g} - \frac{g}{h + g} \ln \frac{g}{h + g}] \ . \tag{27.5}$$

We now know that the ENTROPY of the probability DISTRIBUTION is

$$e(p_1, p_2, \ldots, p_n) = -\sum p_i \ln p_i \ ,$$

$p_i \geq 0$, $\sum p_i = 1$, is largest for a uniform distribution $p_1 = \ldots = p_n$.

The cost function $l(x)$, therefore, reaches its maximum value with fixed σ if

$$\frac{h}{h + g} = \frac{g}{h + g}$$

i.e., for $h = g$. In general, one can say:

The expected value of the cost $l(x)$ increases, if $h \to g$ for a fixed $h + g$.

(27.6)

Furthermore, for $g \geq h$

$$\frac{\partial l}{\partial h} > 0 \quad \text{and} \quad \frac{\partial l}{\partial g} > 0 \ .$$

We now show $\frac{\partial l}{\partial h} > 0$ by differentiating (27.5).

$$\frac{\partial}{\partial h} l = \frac{\sigma}{m} \left[-\frac{h}{h+g} \ln \frac{h}{h+g} - \frac{g}{h+g} \ln \frac{g}{h+g} \right]$$

$$+ (h+g) \frac{\sigma}{m} \left[-\frac{g}{(h+g)^2} \ln \frac{h}{h+g} + \frac{g}{(h+g)^2} \ln \frac{g}{h+g} \right]$$

$$= \frac{\sigma}{m} \left[\underbrace{\qquad\qquad}_{> 0} \right]$$

$$+ \underbrace{\frac{\sigma}{m} \frac{g}{h+g} \ln \frac{g}{h}}_{\geq 0} .$$

Since the function l(x) remains unchanged when exchanging h and g, it also follows that

$$\frac{\partial l}{\partial g} > 0.$$

Here is an example. Let $\sigma = 1$ and
a) $h = g = 1$;
b) $h = 0.1; g = 10$

In both cases, the geometric mean of h and g are equal to one, but
a) $l(x) = 0.77$;
b) $l(x) = 0.317$

$h \neq g$ means that there is a favorable and an unfavorable stock supply for the single period model. The result (27.6) in the above investigation asserts: the more h and g differ, the greater is the cost reduction by applying an optimal ordering rule. This is valid for any demand distribution.

§28   TEMPORAL STRUCTURE OF THE NEWSBOY PROBLEM

Optimal Period Length

Instead of the newsboy we now consider an ice cream vendor in a football stadium. He may only sell ice cream inside the stadium. He sells until his stock have been sold out or until the end of the game. It is assumed that he does not replenish his stock of ice cream. Since he can freely choose when to begin selling, he (in a way) freely chooses the length of the selling period. In his case, is there an optimal period length in this single period problem?

As before, we had assumed a Poisson demand which we then approximated using the logistic function. With a Poisson process, the expected value and variance are proportional to time (compare §19)

$$\mu_T = \sigma_T^2 = \lambda T, \quad \text{i.e. } \sigma_T = \sigma_0 \sqrt{T} .$$

The holding and shortage costs are also proportional to T

$$h_T = hT; \quad g_T = gT.$$

With this, the time–dependent expression for the single period cost with logistically distributed demand (27.5) is given as

$$l_T(x) = (h + g)T \frac{\sigma_0}{m} \sqrt{T} \exp\left[-\frac{h}{h+g} \ln \frac{h}{h+g} - \frac{g}{h+g} \ln \frac{g}{h+g}\right] .$$

The expected total costs per time are

$$C = \frac{l_T(x)}{T} + \frac{k}{T} = (h + g) \frac{\sigma_0}{m} \sqrt{T} \exp\left[-\frac{h}{h+g} \ln \frac{h}{h+g} - \frac{g}{h+g} \ln \frac{g}{h+g}\right] + \frac{k}{T}.$$

The optimal period length with logistically distributed demand is

$$T^* = \left[\frac{2k}{(h_0 + g_0)\frac{\sigma_0}{m} \exp\left[-\frac{h}{h+g} \ln \frac{h}{h+g} - \frac{g}{h+g} \ln \frac{g}{h+g}\right]} \right]^{\frac{2}{3}} .$$

A rough simplification is, however, assumed here: The random result "a demand occurs" is set at exactly the end of the period.

It would be more precise to consider the time of occurrence of the demand within the period.

A More Accurate Formulation

We again assume a Poisson demand. In the model with a fixed period length (exactly one time unit)

$$l(x) = (h + g) \sum_{u=0}^{x} P(u) + g(\mu - x) .$$

Now $P(u) = P_t(u)$ and $\mu = \mu_t$, and the cost function is

$$l_T(x) = \int_0^T \{(h + g) \sum_{u=0}^{x} P_t(u) + g(\mu_t - x)\} dt . \tag{28.1}$$

For a Poisson demand with rate λ

$$P_T(u) = \sum_{j=0}^{u} \frac{(\lambda t)^j}{j!} e^{-\lambda t} ;$$

$$\mu_t = \lambda t .$$

With these expressions, the objective function (28.1) becomes

$$l_T(x) = \int_0^T \left\{ (h + g) \sum_{u=0}^x \sum_{j=0}^u \frac{(\lambda t)^j}{j!} e^{-\lambda t} \right\} dt + g \int_0^T \lambda t \, dt - g \, x \, T \, . \tag{28.2}$$

Intermediate computation:

$$\int_0^T \frac{(\lambda t)^j}{j!} e^{-\lambda t} \, dt = -\frac{1}{\lambda} \frac{(\lambda T)^j}{j!} e^{-\lambda T} + \int_0^T \frac{(\lambda t)^{j-1}}{(j-1)!} e^{-\lambda t} \, dt =$$

$$\underbrace{}_{u} \quad \underbrace{}_{v'} \quad = \dots \text{ (partial integration continued) } \dots =$$

$$= \frac{1}{\lambda} \left[1 - \sum_{r=0}^j \frac{(\lambda T)^r}{r!} e^{-\lambda T} \right] =$$

$$= \frac{1}{\lambda} \left[1 - P_T(j) \right] \, .$$

With the help of this intermediate computation, (28.2) becomes

$$l_T(x) = \frac{(h + g)}{\lambda} \sum_{u=0}^x \sum_{j=0}^u \left[1 - P_T(j) \right] + g\lambda \frac{T^2}{2} - g \, x \, T \tag{28.3}$$

Approximation

We approximate the Poisson distribution for large λT by the logistic

$$P_T(u) = \frac{1}{1 + e^{-\frac{m}{\sigma_T}(u - \mu_T)}} \, . \tag{28.4}$$

From (28.3), we derive

$$l_T(x) = \frac{h + g}{\lambda} \int_{u=0}^{x} \int_{r=0}^{u} \left[1 - \frac{1}{1 + e^{-\frac{m}{\sigma_T}(r - \mu_T)}} \right] dr \, du + g\lambda \frac{T^2}{2} - g \, x \, T \ .$$

$$(28.5)$$

We substitute

$$\int_{-\infty}^{u} \left[1 - \frac{e^{\frac{m}{\sigma}(r - \mu_T)}}{1 + e^{\frac{m}{\sigma}(r - \mu_T)}} \right] dr = u - \frac{\sigma}{m} \ln \left[1 + e^{\frac{m}{\sigma}(u - \mu_T)} \right]$$

in (28.5) and obtain

$$l_T(x) = \frac{h + g}{\lambda} \int_{u=0}^{x} \left\{ u - \frac{\sigma}{m} \ln \left[1 + e^{\frac{m}{\sigma}(u - \mu_T)} \right] \right\} du + g\lambda \frac{T^2}{2} - g \, x \, T.$$

$$(28.6)$$

This cost function is now minimized with respect to x (for a fixed T).

$\frac{dl}{dx} \overset{!}{=} 0$ gives

$$\frac{h + g}{\lambda} \left[x - \frac{\sigma}{m} \ln \left[1 + e^{\frac{m}{\sigma}(x - \mu_T)} \right] \right] - gT = 0 \ ;$$

$$\frac{g\lambda T}{h + g} = x - \frac{\sigma}{m} \ln \left[1 + e^{\frac{m}{\sigma}(x - \mu_T)} \right] =$$

$$= x - \frac{\sigma}{m} \ln \left\{ \left[e^{-\frac{m}{\sigma}(x - \mu_T)} + 1 \right] e^{\frac{m}{\sigma}(x - \mu_T)} \right\} =$$

$$= -\frac{\sigma}{m} \ln \left[1 + e^{-\frac{m}{\sigma}(x - \mu_T)} \right] + \mu_T \ .$$

Since $\mu_T = \lambda T$ and $\sigma = \sqrt{\lambda T}$, therefore

$$\lambda T \frac{h}{h + g} = \frac{\sqrt{\lambda T}}{m} \ln \left[1 + e^{-\frac{m}{\sqrt{\lambda T}}(x - \lambda T)} \right]$$

The solution to x gives the optimal lot size

$$x = \frac{\sqrt{\lambda T}}{m} \ln \frac{1}{e^{m\sqrt{\lambda T} \cdot \frac{1}{1 + g/h}} - 1} + \lambda T \qquad (28.7)$$

A feasibility consideration shows:

as g/h increases, the optimal lot size x also increases.

One can also approximate the Poisson distribution by the normal distribution in the objective function (28.2)

$$\sum_{j=0}^{u} \frac{(\lambda t)^j}{j!} e^{-\lambda t} \approx N(\frac{u - \mu}{\sigma})$$

with $\mu = \lambda t$; $\sigma = \sqrt{\lambda T}$. The optimal lot size x, however, can no longer be explicitly specified.

§29   EXACT FORMULATION

We now want to derive the exact formulation for a Poisson demand. As before, let

u :   Demand within T

$p_u(T)$: Probability that u pieces are demanded in [o,T]

x :   Starting inventory

$$p_u(T) = \frac{(\lambda T)^u}{u!} e^{-\lambda T}$$

Two cases occur: $u \leq x$ and $u > x$

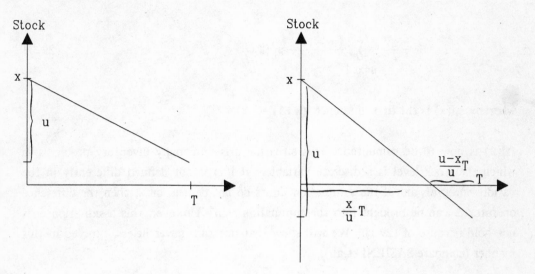

Figure 29.1: Inventory positions in two cases

The holding and shortage costs during period T are formulated as

$$
f_u(x) = \begin{cases} h \cdot T \cdot \dfrac{x + x - u}{2} , & \text{for } u \leq x , \\[3mm] \dfrac{hx}{2} \cdot \dfrac{x}{u} T + \dfrac{g(u-x)}{2} \cdot \dfrac{u - x}{u} \cdot T , & \text{for } u > x . \end{cases}
\tag{29.1}
$$

The expected value of this single period cost with starting inventory x is

$$
1(x) = \sum_{u=0}^{\infty} f_u(x) \frac{(\lambda T)^u}{u!} e^{-\lambda T}
\tag{29.2}
$$

$$l(x) = hT \sum_{u=0}^{x} \left(x - \frac{u}{2}\right) p_u(T) + \frac{hT}{2} x^2 \sum_{u=x+1}^{\infty} \frac{p_u(T)}{u}$$

$$+ \frac{gT}{2} \sum_{u=x+1}^{\infty} \frac{(u - x)^2}{u} p_u(T) \, , \tag{29.3}$$

whereby $\Delta l(x)$ is the first difference $l(x+1) - l(x)$.

$\Delta l(x)$ is now to be computed. This situation arises in many inventory problems in which the stock level is a discrete variable. If $l(x)$ is not defined differently in the various intervals and if the summation limits do not depend on x, then the difference operator Δ can be brought into the summation sign. However, this assumption does not hold because of (29.1). We will show that one can, nevertheless, proceed in this manner (compare SASIENI et.al.).

The holding and shortage cost function, f, is piecewise composed of two partial functions f_1 and f_2 for all defined values of x

$$f_u(x) = \begin{cases} f_{1,u}(x) & , \text{ for } u \leq x \, , \\ f_{2,u}(x) & , \text{ for } u > x \, . \end{cases}$$

Let

$$\tilde{f}_u(x) = f_u(x) \frac{(\lambda T)^u}{u!} e^{-\lambda T}$$

then (29.2) can be written as

$$l(x) = \sum_{u=0}^{\infty} \tilde{f}_u(x) \, .$$

It now applies that for any monotonicly increasing summation limits $a(x)$ and $b(x)$

$$
l(x+1) = \sum_{u=a(x+1)}^{b(x+1)} \tilde{f}_u(x+1)
$$

$$
= \sum_{a(x)}^{b(x)} \tilde{f}_u(x+1) + \sum_{b(x)+1}^{b(x+1)} \tilde{f}_u(x+1) - \sum_{a(x)}^{a(x+1)-1} \tilde{f}_u(x+1) ,
$$

and hence

$$
\Delta l(x) = \sum_{a(x)}^{b(x)} \Delta \tilde{f}_u(x) + \sum_{b(x)+1}^{b(x+1)} \tilde{f}_u(x+1) - \sum_{a(x)}^{a(x+1)-1} \tilde{f}_u(x+1) . \qquad (29.4)
$$

Since

$$
\tilde{f}_u(x) = \begin{cases} \tilde{f}_{1,u}(x) , & \text{for } u \le x , \\[2mm] \tilde{f}_{2,u}(x) , & \text{for } u > x , \end{cases}
$$

then

$$
l(x) = \sum_{u=0}^{b(x)} \tilde{f}_{1,u}(x) + \sum_{u=b(x)+1}^{\infty} \tilde{f}_{2,u}(x) ,
$$

whereby $b(x) = x$.

To determine Δl we now turn to (29.4) for both sums on the right hand side and obtain

$$
\Delta l(x) = \sum_{u=0}^{b(x)} \Delta \tilde{f}_{1,u}(x) + \sum_{u=b(x)+1}^{\infty} \Delta \tilde{f}_{2,u}(x) + \sum_{b(x)+1}^{b(x+1)} [\tilde{f}_{1,\,u}(x+1) - \tilde{f}_{2,\,u}(x+1)]
$$

Since $b(x) = x$, the last summation is limited to

$$\tilde{f}_{1,x+1}(x+1) - \tilde{f}_{2,x+1}(x+1) \ .$$

It has the value zero, since one knows from (29.1) that the equation $f_{1,u}(x) = f_{2,u}(x)$ applies for $u = x + 1$. One can therefore bring the difference operator into the summation sign.

The optimality condition for this discrete problem is

$$\Delta l(x-1) < 0 < \Delta l(x) \ .$$

This leads to

$$\Delta l(x) = (h + g)T \left\{ \sum_{u=0}^{x} p_u(T) + (x + \tfrac{1}{2}) \sum_{u=x+1}^{\infty} \frac{p_u(T)}{u} \right\} - gT \ . \qquad (29.5)$$

The minimization of the expected single period cost means:
Choose the smallest integer x which satisfies the condition

$$\boxed{M(x) > \frac{g}{h + g}} \qquad , \qquad (29.6)$$

whereby

$$M(x) = \sum_{u=0}^{x} p_u(T) + (x + \tfrac{1}{2}) \sum_{u=x+1}^{\infty} \frac{p_u(T)}{u}$$

and, especially for a Poisson demand,

$$M(x) = \sum_{u=0}^{x} \frac{(\lambda T)^u}{u!} e^{-\lambda T} + (x + \tfrac{1}{2}) \sum_{u=x+1}^{\infty} \frac{(\lambda T)^u}{u \cdot u!} e^{-\lambda T} \ . \tag{29.7}$$

If one wants to determine the value of the objective function $l(x)$ aside from the optimal lot size, then one starts at best with $k=0$ and computes the series according to the value $M(k)$ for $k = 1, 2, \dots$ until the condition (29.6) is satisfied for the first time. The corresponding k is the optimal lot size x. One uses the values $M(k)$ for the computation of $l(x)$.

From

$$\Delta l(x) = (h + g) TM(x) - gT \ .$$

one easily obtains $l(x)$:

$$l(x) = l(0) + \sum_{k=0}^{x-1} \Delta l(k) \ .$$

Since

$$l(0) = \frac{gT}{2} \sum_{u=0}^{\infty} u p_u(T) = \frac{gT}{2} E\{u\} \overset{\text{(Poisson)}}{=} \frac{gT}{2} \lambda T \ ,$$

then

$$l(x) = \frac{g \lambda T^2}{2} + \sum_{k=0}^{x-1} \Delta l(k) . \tag{29.8}$$

This is the expected value of the inventory and shortage costs for a period of length T.

Optimal Period Length

Until now the period length T was fixed. We now approximately compute the minimum average cost of a period per unit time.

$$\text{Min}_{T} \; c(T) = \text{Min}_{T} \left\{ \frac{k}{T} + \frac{l_T(x)}{T} \right\} \tag{29.9}$$

Simplest path: $c(T)$ is a convex function with $\lim_{T \to \infty} c(T) = \infty$.

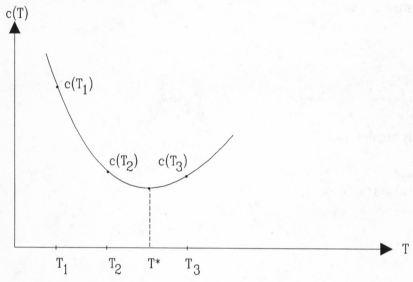

Figure 29.2: Optimal period length

We compute for three different values T_1, T_2, T_3 which should lie near T^*, the average costs $c(T_1)$, $c(T_2)$, $c(T_3)$ and approximate $c(T)$ by the function of type $f(T) = \frac{\alpha}{T} + \beta + \gamma \cdot T$.

This is clearly determined by the three points $(T_1, c(T_1))$, $(T_2, c(T_2))$, $(T_3, c(T_3))$. The minimum lies at

$$\boxed{T^* = \sqrt{\frac{\alpha}{\gamma}}} \; . \tag{29.10}$$

§30 OVERBOOKING

A standard example for overbooking is hotel reservations: A conference is being held in a large hotel during the peak tourist season. The visitors register their participation with the organizer. The organizer, in turn, negotiates a price discount with the hotel manager and books overnight stays for the registered participants.

The hotel manager knows from experience that for large conferences a number of registered participants fail to arrive without cancelling their reservations (so–called "no shows"). It may be profitable for him to keep fewer rooms reserved than were booked. Let

b :   Booked rooms (each participant books a single room)

x :   Vacant rooms (capacity)

h :   Reservation cost of a room for a "no show". A participant who is not coming pays only the discounted price. Had one known that he would not come, the room would have been rented at the normal price. h is, at the same time, the day's discount (normal price minus the discounted price).

g :   Shortage cost. The participant whose room is overbooked arrives. The hotel must assume the cost of the external accomodation of the guest at a higher price category.

u :   Number of participants who have reserved rooms and who actually arrive

q :   Probability of a "no show"

Given b bookings the probability that u guests actually arrive is

$$p_{u;b} = \binom{b}{u}(1-q)^u q^{1-u} \tag{30.1}$$

and the accumulated probability $P(u;b)$ = probability, demand $\leq u$ with b bookings

$$P(u;b) = \sum_{y=0}^{u} \binom{b}{y}(1-q)^y q^{1-y} \ . \tag{30.2}$$

This optimization problem is a newsboy–type problem. The decision variable x is the number of rooms reserved for the conference (inventory). The optimal inventory according to (26.3) is

$$x = P^{-1}\left(\frac{g}{h + g}\right) . \tag{30.3}$$

The binomial distribution defined above has the expected value and variance

$$\mu = b(1 - q) ; \quad \sigma^2 = bq(1 - q) .$$

If b is large, one approximates this distribution by the normal distribution (so–called normal approximation). Then from (30.3)

$$N\left[\frac{x - b(1 - q)}{\sqrt{bq(1 - q)}}\right] = \frac{g}{g + h} .$$

N is the distribution function of the standardized normal distribution. If one approximates the normal distribution by the logistic, one obtains from the above relation

$$\frac{1}{1 + e^{\frac{-m}{\sqrt{bq(1-q)}}[x - b(1-q)]}} = \frac{1}{1 + \frac{h}{g}}$$

$$e^{\frac{-m}{\sqrt{bq(1-q)}}[x - b(1-q)]} = \frac{h}{g} .$$

We solve this equation for x and obtain the following equation for the optimal lot size

$$\boxed{x = \frac{\sqrt{bq(1-q)}}{m} \ln\frac{g}{h} + b(1 - q)} .$$

Again

$$x \left\{ \begin{matrix} \geq \\ < \end{matrix} \right\} \mu \Leftrightarrow \frac{g}{h} \left\{ \begin{matrix} \geq \\ < \end{matrix} \right\} 1 \,,$$

also applies, i.e., if the shortage cost is larger, inventory is larger than the expected utilization. It is the other way around if the holding cost is larger than the shortage cost.

CHAPTER 4:
STOCHASTIC MODELS
WITH CONTINUOUS REVIEW

§31   METHOD OF STATE PROBABILITIES

We already encountered in §23 an inventory model with continuous review. A Poisson demand was assumed. It was shown that under this special assumption the optimal order quantity D is the same as that in the deterministic model with a constant demand rate. D was obtained using the Wilson formula. The interpretation of the objective function C in the stochastic sense led to the method of state probabilities.

In this chapter, we want to generalize the model with continuous review with respect to the demand process and the delivery schedule . In this case, we again use, among others, the model of state probabilities. The basic idea of this method can be summarized in three steps.

Step 1:   Determination of the structure of the optimal ordering rule in parametric form (for example, "order D, if y = 0"; D is the parameter with a still unknown optimal value).

Step 2:   Derivation of the stationary state probabilities. Let $\pi_y(t)$ be the probability that the system is in state y at time t. This means

$$\pi_y^{(D)} := \lim_{t \to \infty} \pi_y^{(D)}(t)$$

is the probability of the state y when using the fixed lot size D.

In general, the stationary distribution $\pi^{(D)}$ depends on the initial distribution $\pi^{(D)}(0)$ and on the parameters D of the ordering rule. It can be shown that the distribution $\pi^{(D)}$ exists for the present inventory model and the ordering rule "order D in case y = 0" is independent of the starting distribution $\pi^{(D)}(0)$.

Step 3:   Minimization of the expected costs per unit time, that is the stationary expected value

$$\sum_y C_y \tau_y^{(D)} \to \underset{D}{\text{Min}} \ . \tag{31.1}$$

C_y :   Cost per unit time in state y

In the case of a Poisson demand, inventory is uniformly distributed at any randomly selected time (compare §23).

We now generalize the demand process. We assume that purchases come one after the other (rather than in batches). Let

p_u :   Probability that a customer buys u units, u = 0,1,2,...

We also assume a Poisson process for customer arrivals . Hence, the demand process is described by a compound Poisson process (compare §19).

A time–dependent consideration, i.e., a cost recursion t → t + Δt, is complicated. Since the objective function depends only on the expected value, one can assume that the customer arrivals are $1/\lambda$ time units apart. This is the mean value of an intermediate arrival interval. It simplifies the stochastic process into a Markov chain whereby at each event a transition from one inventory state to another (in the same state if u = 0) takes place.

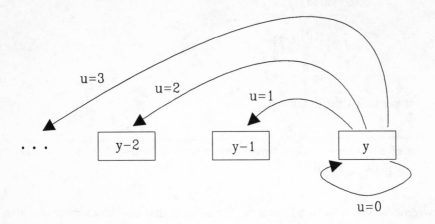

Figure 31.1: State transition diagram

If demand is greater than the stock then the new state is y = 0 and the unsatisfied demand is lost.

For the ordering rule we again define the known structure as

$$\text{Order Quantity } z(y) = \begin{cases} 0, \text{ for } y > 0\,, \\ D, \text{ for } y = 0\,. \end{cases}$$

In this Markov chain model, the transitions take place after each $1/\lambda$ time unit. For the state y = 0, the following conditions apply: the system persists $1/\lambda$ time units in this state and the order is only placed at the end of the period (for immediate delivery!). Thus no inventory costs occur during this period.

The equations which determine the stationary state probabilities are:

$$\pi_D = \sum_{i=0}^{D} \pi_i \sum_{u=i}^{\infty} p_u \; ; \qquad\qquad (31.2)$$

$$\pi_y = \sum_{i=y}^{D} \pi_i p_{i-y} \; ; \qquad 0 \le y \le D-1 \qquad\qquad (31.3)$$

$$\sum_{i=0}^{D} \pi_i = 1 \qquad \text{(Normalizing Equation)} \qquad\qquad (31.4)$$

There are $D+2$ equations with $D+1$ unknowns; however, one equation is linearly dependent, since (31.2) and (31.3) only determine the values π_y, $y = 0, 1, 2, \ldots D$ relative to each other. Therefore one still needs the normalizing equation (31.4).

Geometric Distribution of Demand

Let the demand u of a customer be geometrically distributed

$$p_u = (1-p)p^u; \quad 0 < p < 1 \;, \quad u = 0,1,2,\ldots \; .$$

Then

$$\pi_D = (1-p) \sum_{i=0}^{D} \pi_i \sum_{u=i}^{\infty} p^u$$

$$= (1-p) \sum_{i=0}^{D} \pi_i p^i \sum_{u=0}^{\infty} p^u \qquad \Rightarrow \quad \boxed{\pi_D = \sum_{i=0}^{D} \pi_i p^i}$$

$$\pi_y = (1-p) \sum_{i=y}^{D} \pi_i p^{i-y}$$

$$= (1-p)\pi_y + p(1-p) \underbrace{\sum_{i=y+1}^{D} \pi_i p^{i-(y+1)}}_{= \pi_{y+1}} \qquad \Rightarrow \boxed{\pi_y = \pi_{y+1}}$$

$$y = 0,1,\ldots,D-2$$

$$\pi_0 = (1-p) \underbrace{\sum_{i=0}^{D} \pi_i p^i}_{= \pi_D} \qquad\qquad \Rightarrow \boxed{\pi_0 = (1-p)\pi_D} \quad .$$

As a whole:

$$\pi_y = \begin{cases} \pi_D & , \text{ for } y = D, \\ (1-p)\pi_D, & \text{ for } 0 \le y \le D-1. \end{cases}$$

With the help of the normalization condition (31.4), π_D can now be calculated

$$\sum_{i=0}^{D} \pi_i = 1 \quad \Rightarrow \quad \pi_D = \frac{1}{1 + D - Dp} \quad .$$

The stationary state probabilities, therefore, are

$$\pi_y = \begin{cases} \dfrac{1}{1 + D - Dp} & , \text{ for } y = D , \\[3mm] \dfrac{1-p}{1 + D - Dp} & , \text{ for } 0 \le y \le D-1 . \end{cases} \qquad (31.5)$$

The objective function dependent on D can now be formulated. The costs C_y in state y are

$$C_y = \begin{cases} hy & \text{, for } 1 \leq y \leq D , \\ k + aD & \text{, for } y = 0 . \end{cases} \qquad (31.6)$$

The expected cost per unit time (31.1) in this case is

$$hD\pi_D + h \sum_{y=1}^{D-1} y\pi_y + (k + aD)\pi_0 \to \underset{D}{\text{Min}} .$$

If one substitutes for the state probabilities the values found in (31.5) the following objective function is obtained

$$\boxed{\frac{k + aD + h\frac{D(D-1)}{2}}{\frac{1}{1-p} + D} + \frac{hD}{1 + D(1-p)} \to \underset{D}{\text{Min}}} \qquad (31.7)$$

The first fraction has a similarity to the objective function (2.1) of the Wilson model. From (31.7) no explcit formula can be derived for the optimal lot size D^*. The objective function (31.7), however, can be easily evaluated. The recommended method is to start the evaluation with the discrete Wilson lot size and to continue the computation within a discrete neighborhood until one has found the minimal value.

§32   POISSON DEMAND, EXPONENTIAL DELIVERY TIME

Now we consider models with delivery times.

Let the demand be Poisson distributed and the delivery period exponentially distributed.

λ :   Demand rate
μ :   Delivery rate

Since the delivery period is greater than 0, it will never be optimal to order when y = 0. An order is placed when y = s > 0.

Let the lot size be D. Then the value

$$S = s + D$$

is the maximum inventory from a given period up to the time when the stock first reaches the point s.

Since the stock is continuously reviewed, an order is placed exactly at y = s. Inventory can be allowed to drop until it reaches this value again. However, another order is not allowed until the last order has been used up.

The ordering rule is a type of the so—called

(s,D)—policy

also known as the Two—Bin Policy.

It was practiced in the past by herring sellers. They had an open barrel and, in addition, a closed barrel in reserve. As soon as the open barrel is emptied, the second barrel is opened and a new barrel is immediately ordered.

For models with delivery time,

$$D \geq s \tag{32.1}$$

is more reasonable. If D < s and the stock is allowed to drop to y = 0 before the arrival of the delivery, then the new stock after the arrival of the delivery would be y = D < s and one must immediately order again.

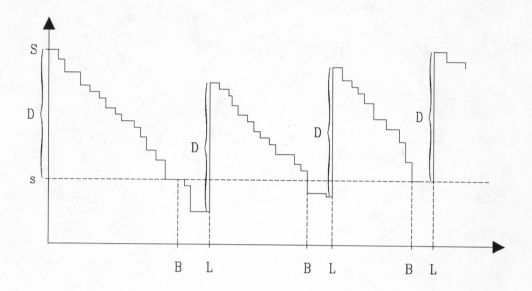

Figure 32.1: Operational characteristic of inventory for a (s,D)–policy.
B = order; L = delivery; L – B = delivery time

Sales are lost as a result of stock deficits (Lost Sales Case).

How large are the state probabilities in this model? We consider the following cases:

Case 1: y = 0

The state y = 0 takes a special position as a boundary point of the state space. The state transition diagram related to a small time period Δt looks as follows:

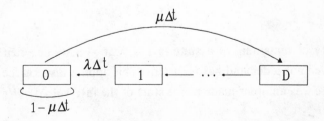

Figure 32.2

The arrows are the transitions probabilities. The probability of remaining in state $y=0$ or arriving in it after a small time span Δt is

$$\pi_0(t + \Delta t) = [1 - \mu\Delta t]\pi_0(t) + \lambda\Delta t \pi_1(t) \ . \tag{32.2}$$

As $\Delta t \to 0$, it becomes

$$\frac{d\pi_0(t)}{dt} = -\mu\pi_0(t) + \lambda\pi_1(t) \ . \tag{32.3}$$

In the stationary case, $\lim\limits_{t \to \infty} \dot{\pi}_0(t) = 0$, i.e.

$$\pi_1 = \frac{\mu}{\lambda}\pi_0 \ . \tag{32.4}$$

Case 2: $1 \leq y \leq s$.
The state transition diagram has the form

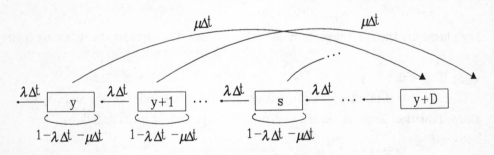

Figure 32.3

The probability of remaining in a state is $1 - \lambda\Delta t - \mu\Delta t$, i.e., neither the existing order has arrived nor a demand has occurred. This applies also for the state $y = s$. At the latest, there was an order made at the start of the interval Δt. We have

$$\pi_y(t + \Delta t) = [1 - \lambda\Delta t - \mu\Delta t]\pi_y(t) + \lambda\Delta t \pi_{y+1}(t) \ . \tag{32.5}$$

In the stationary case

$$\pi_{y+1} = \frac{\lambda + \mu}{\lambda}\,\pi_y \,, \quad 1 \leq y \leq s \,. \tag{32.6}$$

Case 3: $s < y < D$

These states can also be reached only from higher stock levels, as the state–probability diagram shows

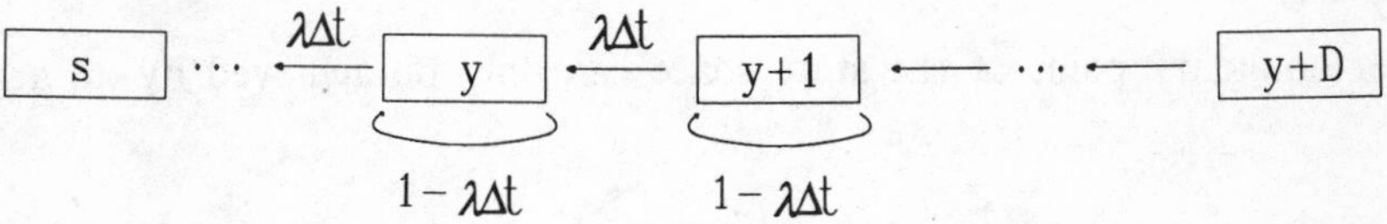

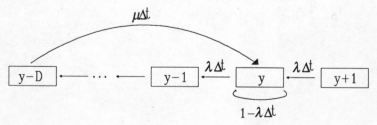

Figure 32.4

The recursive equation of the state probabilites is

$$\pi_y(t + \Delta t) = [1 - \lambda\Delta t]\,\pi_y(t) + \lambda\Delta t\,\pi_{y+1}(t) \,. \tag{32.7}$$

The stationary solution is

$$\pi_{y+1} = \pi_y \,. \tag{32.8}$$

Case 4: $D \leq y < S$

These states occur as a result of a demand as well as a delivery

Figure 32.5

Then

$$\pi_y(t + \Delta t) = [1 - \lambda \Delta t]\pi_y(t) + \lambda \Delta t \pi_{y+1}(t) + \mu \Delta t \pi_{y-D}(t) , \qquad (32.9)$$

from which follows

$$\pi_y = \pi_{y+1} + \frac{\mu}{\lambda}\pi_{y-D} . \qquad (32.10)$$

<u>Case 5:</u> $y = S$

The upper boundary point of the state space can only be achieved by an arrival of goods.

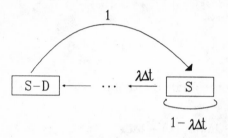

Figure 32.6

We have

$$\pi_S(t + \Delta t) = [1 - \lambda \Delta t]\pi_S(t) + \mu \Delta t \pi_{S-D}(t) \qquad (32.11)$$

and

$$\pi_S = \frac{\mu}{\lambda}\pi_{S-D} . \qquad (32.12)$$

In summary,

$$\pi_1 = \frac{\mu}{\lambda} \pi_0 \; ; \qquad\qquad y = 0$$

$$\pi_{y+1} = (\frac{\lambda + \mu}{\lambda}) \pi_y \; ; \qquad 0 < y \le s$$

$$\pi_{y+1} = \pi_y \; ; \qquad\qquad s < y < D$$

$$\pi_y = \pi_{y+1} + \frac{\mu}{\lambda} \pi_{y-D} \; ; \qquad D \le y < S$$

$$\pi_S = \frac{\mu}{\lambda} \pi_{S-D} \; ; \qquad\qquad y = S \; .$$

We set

$$\rho = \frac{\lambda}{\mu}$$

$$\alpha = \frac{1 + \rho}{\rho}$$

and represent the state probabilities independent of π_0.

$$\pi_y = \frac{\alpha^{y-1}}{\rho} \pi_0 \; ; \qquad\qquad 0 < y \le s$$

$$\pi_y = \frac{\alpha^s}{\rho} \pi_0 \; ; \qquad\qquad s < y \le D$$

$$\pi_y = \frac{1}{\rho} [\alpha^s - \alpha^{y-D-1}] \pi_0 \; ; \qquad D < y \le S-1$$

$$\pi_S = \frac{\alpha^{s-1}}{\rho^2} \pi_0 \; ; \qquad\qquad y = S \; .$$

(32.13)

π is then determined by the normalization condition $\sum\limits_{y} \pi_y = 1$

$$\pi_0 = \frac{1}{D \frac{\alpha^s}{\rho} + 1} \; .$$

(32.14)

In the next step, the average cost C per unit time in the stationary case will be calculated with the help of the state probabilities

$$C = \pi_0 \lambda g + \pi_{s+1} \lambda [k + aD] + h \sum_{y=1}^{S} y \pi_y \tag{32.15}$$

$$= \pi_0 \lambda g + \pi_0 \frac{a^S}{\rho} \lambda [k + aD] + h \pi_0 \beta \, ,$$

whereby

$$\beta = \frac{a^S}{2\rho} [D^2 + 2Ds - 2s - D] + a^S [a(s + D) - s - 2D] + D. \tag{32.16}$$

Substituting π_0 from (32.14) results in

$$C = \frac{k\lambda}{D} + a\lambda + \frac{(g - a)\lambda - \frac{k\lambda}{D} + \beta h}{D \frac{a^S}{\rho} + 1} \, . \tag{32.17}$$

We now attempt to determine the optimal values s^*, D^* . This is still possible in the border line case $u \gg \lambda$.

Borderline Case: $\mu \gg \lambda$

$\rho \ll 1$ follows from $\mu \gg \lambda$ and from it $\alpha \gg 1$. The objective function (32.17) becomes

$$C \to \frac{k\lambda}{D} + a\lambda + \frac{h}{D} \left\{ \frac{1}{2}[D^2 + 2Ds - 2s - D] + \frac{a(s + D) - s - 2D}{(\alpha - 1)} \right\}$$

$$\lim_{\alpha \to \infty} C_\alpha = C^* = \frac{k\lambda}{D} + a\lambda + \frac{h}{2}(D + 2s + 1) \, . \tag{32.18}$$

The equations

$$\frac{\partial C^*}{\partial D} \overset{!}{=} 0 : \qquad \boxed{D = \sqrt{\frac{2k\lambda}{h}} \; ;} \qquad\qquad (32.19)$$

$$\frac{\partial C^*}{\partial s} > 0 : \qquad \boxed{s = 0 \; .} \qquad\qquad\qquad (32.20)$$

give the necessary conditions for an optimum.

We have a boundary extremum with regards to s. As expected, we obtain in the borderline case $\mu >> \lambda$ the results of the model without delivery time from §22.

Reserved Storage Area

In all other cases, one must determine the solutions to D and s either by numerical methods or by simplifying the model such that it can be solved by analytical methods. The source of the difficulties is the term β.

The average cost C in (32.15) depends on all inventory levels $y = 0, 1, 2, \ldots S$. The problem eventually becomes simpler if the inventory cost is measured at its maximum level: $h(s + D)$. This is the case, for example, if one does not own the warehouse, but reserves storage area at an external warehouse. It must be big enough to cover the maximum stock level. The objective function is then

$$C = \pi_0 g\lambda + \pi_{s+1}\lambda[k + aD] + h(s + D) \; .$$

After a short intermediate calculation one obtains

$$\boxed{C = \lambda a + \frac{\lambda(g - a)\rho + k\lambda\alpha^S}{\rho + D\alpha^S} + h(s + D) \; .} \qquad (32.21)$$

The term β no longer appears here. With this, the minimization of C relative to s and D becomes simpler. However, we still cannot avoid numerical methods.

§33   POISSON DEMAND, FIXED DELIVERY TIME τ

We consider an inventory model with continuous review, Poisson demand and a fixed delivery time τ. For the formulation of the model we now use Bellman's Principle of Optimality. Future costs are not discounted.

If we observe the inventory level y in period t, we can influence the level with an immediate action at the earliest period starting from period $t + \tau$. Until then we have no influence on what has already occurred. We apply the cost l(y) which corresponds to an event occurring in period $t + \tau$. Previously placed orders may arrive between t and $t+\tau$. The inventory level $y_{t+\tau}$ is therefore dependent on y_t, on the amount on order and on the decision made in period t. Hence we define in the following model the state y

y :   Stock on hand plus on order

Unsatisfied demand is backlogged (BACKORDER CASE)

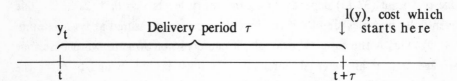

Figure 33.1: Costs are incurred at time $t+\tau$ for a delivery period τ

The inventory level y at time $t + \tau$ is a random variable.

The probability (Inventory = y − u at time $t + \tau$ | Inventory = y at time t)

   = Probability (Demand = u in time τ)

   $= \dfrac{(\lambda\tau)^u}{u!} e^{-\lambda\tau}$ for Poisson Demand.

The expected inventory and shortage costs in period $t + \tau$ are

$$f(y) = h \sum_{u=0}^{y} (y-u) \frac{(\lambda\tau)^u}{u!} e^{-\lambda\tau} + g \sum_{u=y+1}^{\infty} (u-y) \frac{(\lambda\tau)^u}{u!} e^{-\lambda\tau}. \tag{33.1}$$

As was shown in §26 (compare (26.1), (26.2)), this expression may be rewritten as

$$f(y) = (h+g) \sum_{u=0}^{y} P_u + g(\mu - y) ,$$

where

$$P_u = \sum_{i=0}^{u-1} \frac{(\lambda\tau)^i}{i!} e^{-\lambda\tau} .$$

Note: The discrete distribution function is written as $P_x = P(u<x)$ instead of the usual convention $P_x = P(u \leq x)$ so that the above expression (26.2) derived by integration may be used for discrete demand.

To simplify the notation, we define the demand rate λ as demand per unit time τ. Hence, we do not need to use τ explicitly in the notation.

Principle of Optimality
Let

$C:$ Average cost per unit time

For a stationary cost increase, the cost 1 becomes the total cost $1 + C\Delta t$ if one moves the present time period back by Δt. According to the optimality principle of Bellman we have the recursion

$$l(y) + C\Delta t = f(y)\Delta t + [1 - \lambda\Delta t]l(y) + \tag{33.2}$$

$$+ \lambda\Delta t \min_{x \geq y-1} \{k\delta(x - y + 1) + a(x - y + 1) + l(x)\} ,$$

for s < y ≤ S. By cancelling l(y) from both sides and dividing by Δt one obtains

$$\lambda l(y) + C = f(y) + \lambda \min_{x \geq y-1} \{k\delta(x - y + 1) + a(x - y + 1) + l(x)\} \tag{33.3}$$

for s < y ≤ S.

Structure of the Optimal Policy

We begin by considering a starting inventory y = S. This level drops in the course of time. Sooner or later, stocks must again be ordered. This happens for y = s. The order quantity is S − s = D. The structure of the ordering rule is, therefore, of the (s,D)–type. Following this, one can divide (33.3) into

$$\lambda l(S) + C = f(S) + \lambda l(S - 1)$$
$$\lambda l(S - 1) + C = f(S - 1) + \lambda l(S - 2)$$
$$\vdots$$
$$\lambda l(s + 1) + C = f(s + 1) + \lambda [k + aD + l(S)] .$$

The sum of these individual equations gives

$$DC = \sum_{y=s+1}^{S} f(y) + \lambda k + \lambda aD . \tag{33.4}$$

In this case, the average cost C per unit time is a function of the structural parameters s and D. The optimization problem is

$$C = \frac{1}{D} \sum_{y=s+1}^{s+D} f(y) + \frac{\lambda k}{D} + \lambda a \; \rightarrow \; \underset{s,D}{Min} \tag{33.5}$$

The constant λa does not influence s and D. Thus we consider the average cost c, without proportional ordering costs. For simplicity we use the integral representation of the sum

$$c = \left\{ \frac{1}{D} \int_{s}^{s+D} f(x) dx + \frac{\lambda k}{D} \right\} \; \rightarrow \; \underset{s,D}{Min} \tag{33.6}$$

The conditions $\frac{\partial c}{\partial s} = 0$ and $\frac{\partial c}{\partial D} = 0$ necessary for an optimum give

$$f(s + D) = f(s)$$
$$D \cdot f(s + D) - \int_{s}^{s+D} f(x) dx - \lambda k = 0.$$

Substituting the first equation into the second results in

$$(s + D) f(s + D) - sf(s) - \int_{s}^{s+D} f(x) dx = \lambda k$$
$$xf \Big|_{x=s}^{x=s+D} - \int_{s}^{s+D} f(x) dx = \lambda k .$$

The partial integral $\int f \, dx = xf - \int xf' \, dx$ finally leads to the necessary optimal conditions

$$\boxed{\begin{array}{l} \int_{s}^{s+D} xf'(x) dx = \lambda k \\[2mm] f(s + D) = f(s) . \end{array}}$$

$$\tag{33.7}$$
$$\tag{33.8}$$

These two equations can be solved numerically. One may obtain an approximate solution in an analytical manner if one expands both of the above optimality conditions in a Taylor series and solves the system of equations for s and D. One obtains explicit formulas for s and D. The same results, however, can be achieved with less effort if the objective function (33.6) is first approximated by a Taylor series and then the partial derivatives set to zero. It is

$$f(x) = f(\lambda) + (x - \lambda)f'(\lambda) + (x - \lambda)^2 \frac{f''(\lambda)}{2!} + \dots \tag{33.9}$$

Integrating term by term

$$\int_{s}^{s+D} f(x)dx = Df(\lambda) + f'(\lambda\lambda)^2 dx + \dots$$

$$= Df(\lambda) + \frac{f'(\lambda)}{2} D[D + 2(s - \lambda)] +$$

$$+ \frac{f''(\lambda)}{6} D[D^2 + 3D(s - \lambda) + 3(s - \lambda)^2] + \dots ,$$

and stopping after the second order term in x. We obtain the approximate objective function $\tilde{c}$

$$\tilde{c} = \frac{k\lambda}{D} + f(\lambda) + \frac{f'(\lambda)}{2}[D + 2(s - \lambda)] +$$

$$+ \frac{f''(\lambda)}{6}[D^2 + 3D(s - \lambda) + 3(s - \lambda)^2]. \tag{33.10}$$

The condition $\frac{\partial \tilde{c}}{\partial s} = 0$ gives

$$D + 2(s - \lambda) = -2\frac{f'(\lambda)}{f''(\lambda)} . \tag{33.11}$$

The condition $\frac{\partial \tilde{c}}{\partial D} = 0$ results in

$$-\frac{k\lambda}{D^2} + \frac{f'(\lambda)}{2} + D\frac{f''(\lambda)}{3} + \frac{f''(\lambda)}{2}(s - \lambda) = 0. \tag{33.12}$$

If one substitutes the equation (33.11), which was solved for s − λ, into the above equation, one obtains for D an expression independent of s

$$D = \sqrt[3]{\frac{12k\lambda}{f''(\lambda)}}$$

(33.13)

Equation (33.11) gives

$$s = -\frac{f'(\lambda)}{f''(\lambda)} - \frac{D}{2} + \lambda$$

(33.14)

Note: The assumption of a Poisson distribution is implicit in the formulation of the Principle of Optimality (33.2). Since the Poisson distribution is a discrete distribution, one must use the first or the second difference quotient of f instead of its first or second derivative. The approximation by a continuous distribution is simpler.

Approximation by a Normal Distribution

For large λ, the Poisson distribution can be closely approximated by the normal distribution. Let

$N((x-\lambda)/\sqrt{\lambda})$:   Distribution function of the standardized normal distribution, when the Poisson distribution has an expected value of λ.

Then

$$f'(x) = (h + g)N(\frac{x - \lambda}{\sqrt{\lambda}}) - g;$$

$$f'(\lambda) = \frac{h - g}{2};$$

$$f''(\lambda) = \frac{h + g}{\sqrt{\lambda}} \frac{1}{\sqrt{2\pi}};$$

$$D = \sqrt[3]{\frac{12k\sqrt{2\pi}}{h+g}} \sqrt{\lambda} \; ; \qquad\qquad (33.15)$$

$$s = \frac{g-h}{g+h} \cdot \frac{\sqrt{2\pi\lambda}}{2} - \frac{D}{2} + \lambda \; . \qquad\qquad (33.16)$$

Since $\sigma = \sqrt{\lambda}$, it follows from (33.15)

$$D \sim \sigma \; . \qquad\qquad (33.17)$$

Approximation by the Logistic Function

We now approximate the Poisson distribution by the logistic function.

Let

$$L(x;\lambda,\sigma) = \frac{1}{1 + e^{-\frac{m(x-\lambda)}{\sigma}}} \; ; \quad m = \frac{4}{\sqrt{2\pi}} \qquad\qquad (33.18)$$

be the distribution function of a random variable distributed according to the logistic function with expected value λ and standard deviation σ. Then

$$f(x) = (h+g)\frac{\sigma}{m} \ln\left\{1 + e^{\frac{m}{\sigma}(x-\lambda)}\right\} + g(\lambda - x) \; ;$$

$$f'(x) = \frac{h+g}{1 + e^{-\frac{m(x-\lambda)}{\sigma}}} - g \; ;$$

$$f'(\lambda) = \frac{h-g}{2} \; ;$$

$$f''(\lambda) = \frac{h+g}{\sigma}\frac{1}{\sqrt{2\pi}} \; .$$

Since $f'(\lambda)$ and $f''(\lambda)$ are identical for the normal and logistic distributions, we again obtain for D the formula (33.15).

We now substitute $f(x)$ in the necessary optimality condition (33.8) and get

$$(h + g) \, \frac{\sigma}{m} \, \ln \left\{ \frac{1 \, + \, e^{\frac{m}{\sigma}(s+D-\lambda)}}{1 \, + \, e^{\frac{m}{\sigma}(s-\lambda)}} \right\} = gD \; . \tag{33.19}$$

Since D is known from (33.15) one can compute the expression

$$A = e^{\frac{m}{\sigma}\frac{D}{2}} \; .$$

With a further simplification, let

$$V = e^{\frac{m}{\sigma}(s + \frac{D}{2} - \lambda)} \; , \tag{33.20}$$

then equation (33.19) becomes

$$\frac{1 \, + \, A \cdot V}{1 \, + \, \frac{V}{A}} = A^{\frac{2g}{h+g}} =: Z \; ,$$

and it follows that

$$V = \frac{Z \, - \, 1}{A \, - \, \frac{Z}{A}} \; .$$

After computing V, the value of s can be determined. We solve (33.20) for s and obtain

$$\boxed{s = \frac{\sigma}{m} \, \ln V + \lambda - \frac{D}{2}} \tag{33.21}$$

with $\sigma = \sqrt{\lambda}$ and m $= \dfrac{4}{\sqrt{2\pi}}$.

Detailed computations show that equations (33.16) and (33.21) give good approximate values for the optimal value s^*. The approximation given by (33.15) for D, however, results in most cases into an underestimation of the optimal value of D^*.

It is possible, however, to correct for underestimating D. It can be done in the following way:
The function f is expected to assume its minimum near s $+ \dfrac{D}{2}$. However, the value λ, which was expanded around f(x) in a Taylor series (33.9), may be far from the minimum point. In the hope of a better approximation, one can expand f(x) in a Taylor series about the point s $+ \dfrac{D}{2}$ instead of the point λ. The calculation leads, however, to complicated expressions.

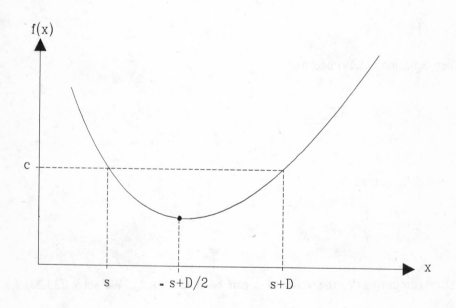

Figure 33.1

But we can improve the value of D by a correction. After D and s have already been computed using (33.15) and (33.16) or (33.21), we can compute an improved D = D_{new} according to the formula

$$D_{new} = \sqrt[3]{\frac{12k\lambda}{f''(s + \frac{D}{2})}}.$$

Cost Function

The following equation applies to the cost function

$$c = \frac{k\lambda}{D} + \frac{1}{D} \int_s^{s+D} f(x)dx =$$

$$= \frac{k\lambda}{D} + \frac{1}{D}[xf(x) \mid_s^{s+D} - \int_s^{s+D} x f'(x)dx] =$$

$$= \frac{k\lambda}{D} + \frac{(s + D)f(s + D) - sf(s)}{D} - \underbrace{\frac{1}{D} \int_s^{s+D} x f'(x)dx}_{= \frac{k\lambda}{D}} \quad (\text{compare } (33.7))$$

With (33.8), it becomes

$$\boxed{c = f(s) = f(s + D)}. \qquad (33.22)$$

§34 POISSON DEMAND, STOCHASTIC DELIVERY TIME, SINGLE ORDER

Under competition conditions, service reliability is an important factor. The supplier will try hard to meet his delivery schedules. In inventory control, the primary variability is in the demand. The opposite is true in a monopolistic situation or in places where goods are rationed; then the primary variability is in the delivery time rather than in demand. This is often observed in developing countries.

We now set up a model with stochastic delivery time. The stock is reviewed continuously. As long as there is still a standing order, no further orders should be placed. Let delivery time and demand be independent of each other. Both follow a Poisson process. This inventory model was already covered in §30. In that section, the method of state probabilities was applied. We attempted to derive a formula for s and D but were not successful. In this section, we apply the Principle of Optimality. Let

$\mu\Delta t$:	Probability that an existing order is delivered in period Δt
$\lambda\Delta t$:	Probability that a unit of good is demanded in period Δt
t :	Time since the last order
t = 0 :	No order is placed.
l(y,t):	Value function in the stationary case

Undiscounted Case

We now formulate the Principle of Optimality for the undiscounted stationary case.

Let t = 0:

$$l(y,0) + C\Delta t = hy\Delta t + [1 - \lambda\Delta t]\, l(y,0) +$$
$$+ \lambda\Delta t\ \underset{D}{\text{Min}}\ \{\text{Min}\ \{k + aD + l(y{-}1,\Delta t)\}\ \mid\ l(y{-}1,0)\},$$

order ———————————————————————⌐

do not order ——————————————————————————⌐

whereby

$$\lambda l(y,0) + C = hy + \lambda\underset{D}{\text{Min}}\ \{\text{Min}\ \{\ .\ \}\ \mid\ l(y{-}1,0)\} \tag{34.1}$$

follows.

With a large initial inventory, an order is not worthwhile. The lower the initial inventory, the lower is the cost advantage of the decision "not to order". At a certain point y = s, it is more favorable to order. Since inventory is reviewed continuously, one immediately orders the quantity D at y = s. Considering this plausibility, the model with delivery time also gives a justification for the (s,D)–policy.

Let t > 0:

As long as there are still pending deliveries, one may not place a new order. This situation creates no room for decision–making. One need not apply the Principle of Optimality. The cost recursion is given by (see §23)

$$1(y,t) + C\Delta t = hy\Delta t + [1 - \lambda\Delta t - \mu\Delta t]1(y,t + \Delta t) + \lambda\Delta t 1(y{-}1,t{+}\Delta t) + \mu\Delta t 1(y{+}D,0).$$

$$(34.2)$$

The boundary condition for y = 0 is given in the lost sales case by

$$1(0,t) + C\Delta t = \lambda\Delta t G + [1 - \lambda\Delta t - \mu\Delta t]\ 1(0,t{+}\Delta t) + \lambda\Delta t 1(0,t{+}\Delta t) + \mu\Delta t 1(D,0).$$

$$(34.3)$$

Let

G : Penalty cost if an order is not delivered thereby causing customer dissatisfaction; this is independent of time (dimension: cost)

As $\Delta t \to 0$, from (34.2), the differential equation becomes

$$-\frac{\partial 1(y,t)}{\partial t} + [\lambda + \mu]\ 1(y,t) = hy + \lambda 1(y{-}1,t) - C + \mu 1(y{+}D,0), \quad t > 0 \qquad (34.4)$$

with boundary condition

$$-\frac{\partial 1(0,t)}{\partial t} = -C + \lambda G - \mu[1(0,t) - 1(D,0)] \ . \tag{34.5}$$

Hence, the optimization problem may be described as a linear differential equation which can be solved by integration.

Discounted Case

We are interested in $1(s,0)$. Let r be the interest rate and e^{-rt} the discount factor from time t to time zero (see §21). With

$q(\tau)d\tau$: Density of the delivery time distribution

one obtains

$$1(s,0) = k + aD + \underbrace{\int_{\tau=0}^{\infty} q(T) \int_{t=0}^{\tau} \sum_{u=0}^{s} f(s-u) \frac{(\lambda t)^u}{u!} e^{-\lambda t} e^{-rt} dt \ d\tau +}_{=: \ F(s)}$$

$$+ \int_0^{\infty} q(\tau) \sum_{u=0}^{s} \frac{(\lambda\tau)^u}{u!} e^{-(\lambda+r)\tau} 1(s - u + D,0) d\tau +$$

$$+ \int_0^{\infty} q(\tau) \sum_{u=s+1}^{\infty} \frac{(\lambda\tau)^u}{u!} e^{-(\lambda+r)\tau} 1(D,0) d\tau \ . \tag{34.6}$$

For $y > s$, the following consideration applies: The average waiting time of the system in state y is $\frac{1}{\lambda}$. For this time, the inventory cost hy is incurred. Inventory then drops to $y - 1$. The resulting costs $1(y-1,0)$ are discounted by the factor $e^{-r\lambda}$

$$1(y,0) = \frac{hy}{\lambda} + \rho 1(y-1,0) \ ; \qquad y > s \ . \tag{34.7}$$

Special Case: Internal Production or Just–In–Time Delivery

If we produce on our own, the supply quantity is under our control. We assume that the good is expedited from the finished goods inventory to the sales warehouse.

Delivery time arises because the order must line up in a queue of already existing delivery orders and, therefore, remains unprocessed for a time. Hence, the amount still to be delivered can be changed at the last minute. In this case, the order quantity can can always be updated up to last moment just before delivery such that inventory is filled up to y = S upon the arrival of an order. Under this assumption, the equation (34.6) changes to

$$1(s,0) = k + aD + F(s) + 1(S,0) \underbrace{\int_0^\infty q(\tau)e^{-r\tau}d\tau}_{=: \ \alpha} \ . \tag{34.8}$$

α :   Expected value of the discount factor over the delivery time

Equation (34.7) remains unchanged. This applies especially for y = S

$$1(S,0) = \frac{h}{\lambda} \sum_{y=s+1}^{S} y + \rho^D 1(s,0) \ . \tag{34.9}$$

Substituting l(s,0) from equation (34.8) and evaluating the sum, one obtains

$$1(S) = \frac{1}{1 - \rho^D \alpha} \left\{ \frac{h}{\lambda} \left[\frac{S(S+1)}{2} - \frac{s(s+1)}{2} \right] + \rho^{S-s}[k + aD + F(s)] \right\} \ . \tag{34.10}$$

The second argument $\tau = 0$ in the cost function l is no longer carried because the recursions (34.9) and (34.10) are always related to $\tau = 0$.

It is also not possible here to express formulas for the optimal values s*, S*. One obtains them by minimizing (34.10)

$$\frac{1}{1 - \rho^D\alpha} \left\{ \frac{h}{\lambda} \left[\frac{S(S+1)}{2} - \frac{s(s+1)}{2} \right] + \rho^{S-s} [k + aD + F(s)] \right\} \to \underset{s,S}{\text{Min}} ,$$

whereby one can restrict s and S to integer values.

§35   POISSON DEMAND, STOCHASTIC DELIVERY TIME, MULTIPLE ORDERS

We now extend the inventory model to the case where a new order may be placed even before an existing order is delivered. We first the treat the case:

Delivery Time τ is Exponentially Distributed

The delivery time distribution has the density

$$q(\tau)d\tau = \mu e^{-\mu\tau} d\tau .$$

The exponential distribution has the advantage that one need not know how old a pending delivery is. We assume that all delivery times are identically distributed (same supplier).

Due to the Poisson demand, all orders are stochastically independent of each other. Because of this reason, the probability of an order occurring is

a)   in case one order is pending: $\mu\Delta t$

b)   in case m orders are pending: $m\mu\Delta t$

The number

m : Pending orders

is the second state variable in this inventory model besides the stock y. Hence, we need not add the pending orders to the inventory.

y : Physical inventory (or shortage)

The inventory and shortage costs are

$$\varphi(y) = \begin{cases} hy, & \text{for } y \geq 0; \\ -gy, & \text{for } y < 0; \end{cases} \quad \text{or} \quad \varphi(y) = \begin{cases} hy, & \text{for } y \geq 0; \\ -G\lambda, & \text{for } y < 0. \end{cases} \tag{35.1}$$

Each single order has a lot size D. The principle of optimality without discounting is given by

$$l(y,m) + C\Delta t = \varphi(y)\Delta t + m\mu\Delta t \, l\,(y+D,m-1) + [1 - m\mu\Delta t - \lambda\Delta t]l(y,m) + $$
$$\lambda\Delta t \, \text{Min } \{k + aD + l(y-1,m+1) \mid l(y-1,m)\} \tag{35.2}$$

or after transposition, simplification and division by Δt

$$(\lambda + m\mu)l(y,m) + C = \varphi(y) + m\mu l(y + D, m - 1) + $$
$$+ \lambda \text{ Min } \{k + aD + l(y-1,m+1) \mid l(y-1,m)\} \ . \tag{35.3}$$

This is a difficult differential equation. We defer, therefore, to a heuristic solution, e.g., in which we introduce equidistant order points s_1, s_2, ...

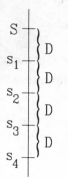

Figure 35.1: A number of equidistant order points

The problem "determine M and D" can be solved with the help of state probabilities.

Delivery Time τ arbitrarily Distributed — No Overlapping

Let the delivery time be arbitrarily distributed. No overlapping means what was ordered first arrives first.

For a fixed delivery time τ we used the concept of relating the inventory cost to the time period τ. We will proceed here in the same manner. Since τ is stochastic, the demand u within τ becomes a random variable. Let

p(u) :   Probability that demand u occurs within the delivery time

For a Poisson demand with rate λ

$$p(u) = \int\limits_{0}^{\infty} \frac{(\lambda\tau)^u}{u!} e^{-\lambda\tau} q(\tau)\, d\tau \; . \tag{35.4}$$

Example: τ is Gamma–distributed

$$q(\tau)d\tau = \frac{\mu^{j+1}\tau^j e^{-\mu\tau}}{j!} d\tau \; ;$$

$$p(u) = \frac{1}{(\lambda+\mu)^{u+j+1}} \int_0^\infty \frac{1}{u!} \frac{1}{j!} \lambda^u \mu^{j+1} (\lambda+\mu)^{u+j+1} \tau^{u+j} e^{-(\lambda+\mu)\tau} d\tau$$

NB: $\int_0^\infty \beta^{i+1} \tau^i e^{-\beta\tau} d\tau = i!$

$$p(u) = \frac{(u+j)!}{u!j!} \left(\frac{\lambda}{\lambda+\mu}\right)^u \left(\frac{\mu}{\lambda+\mu}\right)^{j+1} =$$

$$= \binom{-(j+1)}{u}(-\rho)^u (1-\rho)^{j+1} \; . \tag{35.5}$$

This is a negative binomial distribution with exponent $-(j+1)$ and probability $\rho=\lambda/(\lambda+\mu)$. A Poisson demand occurring within the delivery period, where τ is gamma distributed, becomes an exponentially distributed random value.

The expected value of the inventory and shortage costs $f(y)$ in relation to the expected value of τ is

$$f(y) = h \sum_{u=0}^y (y-u)p(u) + g \sum_{u=y+1}^\infty (u-y)p(u) \; . \tag{35.6}$$

The Principle of Optimality is

$$l(y) + C\Delta t = f(y)\Delta t + [1 - \lambda\Delta t]l(y) +$$

$$+ \lambda\Delta t \min_{x \geq y-1} \{k\delta(x-y+1) + a(x-y+1) + l(x)\} \; . \tag{35.7}$$

It has now become simpler compared to (34.2). It is because all of the remaining orders are ignored. The expected costs depend only on the last order (no overlapping!). However, f has become more complicated. In the limiting case "fixed delivery time", $q(\tau)$ is turned into an uncharacteristic distribution.

Since (35.7) is identical to the principle of optimality (33.2), the minimization of the costs per cycle leads to the same formula as in the model with fixed delivery time (§33) but with another formula (35.5) for $p(u)$.

Example with discounting
We consider in this example the discounted case. Let

r : Interest rate
e^{-rt} : Discount factor
$e^{-r\Delta t} \approx (1 - r\Delta t)$ for $\Delta t << 1$.

In the discounted case, the Principle of Optimality for small Δt is

$$l(y) = f(y)\Delta t + (1 - \lambda\Delta t)(1 - r\Delta t)l(y) +$$

$$+ \lambda\Delta t(1 - r\Delta t) \underset{x\geq y-1}{\text{Min}} \{k\delta(x-y+1) + a(x-y+1) + l(x)\} .$$

If one ignores the terms of higher order in Δt, we have

$$l(y) = \frac{1}{\lambda+r} f(y) +$$
$$\frac{\lambda}{\lambda+r} \underset{x\geq y-1}{\text{Min}} \{k\delta(x - y + 1) + a(x - y + 1) + l(x)\} \qquad (35.8)$$

This formulation is no different from the inventory model with periodic review (compare the Principle of Optimality principle in the formulation (36.4)).

The next chapter is devoted to such models.

CHAPTER 5:
STOCHASTIC MODELS WITH PERIODIC REVIEW

This chapter may be read independently of the previous chapters.

§36   THE ARROW–HARRIS–MARSCHAK MODEL

Continuous inventory monitoring is no longer a problem with the introduction of electronic data processing. Many firms, however, still maintain periodic inspection and decision. Sometimes it is due to arrangements made with suppliers which only allow an order at specific (often equidistant) points in time. Two successive possible points define a period. It is then superfluous to monitor the stock during the period since one does not have any use for this information. An inventory review (physical or book) at the start of each period is enough.

Multi–period models with stochastic demand require a more flexible ordering rule than models with continuous monitoring. The lot size is chosen depending on the actual inventory. Hence, such models require dynamic programming (DP) as a method of solution. For a rigorous handling of this problem, dynamic programming is applied here for the first time. The following is, at the same time, an introduction to the thinking process of dynamic programming. (The Principle of Optimality, from which DP is based, was already formulated a number of times in the previous sections; a computational method of DP was already used in §25.) Some algorithms for stochastic DP are explained in Chapter 6.

The basic model of inventory with periodic monitoring was formulated by KENNETH ARROW, TED HARRIS and JACOB MARSCHAK and is named AHM–Model after them. (ARROW & HARRIS & MARSCHAK (1951)).

The following conditions apply:
1. Periodic review and decision
2. For all periods, demand is random, independent and identically distributed.

3. Unsatisfied demand at the end of each period
  a) is lost (LOST SALES)
  b) is backlogged (BACKORDER)
4. Deliveries arrive immediately. Delivery time is zero.
5. Finite or infinite planning horizon.
6. Discounting

Let

ρ	:	Discount factor for one period
n	:	Planning horizon
y	:	Inventory at the start of a period immediately before a decision
x	:	Inventory at the start of a period immediately after a decision
x–y	:	Order quantity
u	:	Demand, random variable
P(u)	:	Distribution function of demand with density $p_u du$
f(x)	:	Expected value of inventory and shortage costs for a period
a	:	Proportional order cost
k	:	Fixed order cost

The cost function was previously denoted by l. In dynamic programming, the value function is usually denoted by v. Since periodic models play a great role in DP, we assume the following notation (in place of l we now use v):

v	:	Value function (previously l)
v_0	:	Costs incurred at the end of the planning horizon
$v_0 \equiv 0$	:	By assumption (as long as no other value is set)

The following figure shows the operational characteristic of inventory control using the so–called (s,S)–Policy (compare 39.1). The heavy lines show the inventory movement along time; the dashed lines show other possible inventory levels.

The value s is the order limit and S is the full inventory point.

The (s, S) policy states: If $y \leq s$ at the time of inspection, then place an order sufficient to raise inventory up to S. (Delivery times are ignored for a moment.)

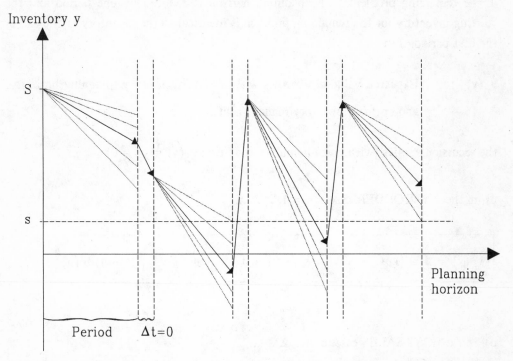

Figure 36.1: Operational characteristic of the AHM Model

The expected inventory and shortage costs with inventory cost rate h and shortage cost rate g are

$$f(x) = h \int_0^x (x-u) \, dP(u) + g \int_x^\infty (u-x) \, dP(u) =$$
$$\overset{(\S 26)}{=} (h+g) \int_0^x P(u) \, du + g(\mu - x) \ . \tag{36.1}$$

μ is the expected value of demand.

According to the BERNOULLI Principle, we choose as our objective function the expected value of all costs during the entire planning horizon. We divide these costs into the costs of the period immmediately prior to it (single period costs) and the costs of the remaining problem. The planning horizon is reduced by one period and the starting inventory for the remaining problem is identical to the inventory at the end of the first period. For

$v_n(y)$:   Expected value of all costs with initial inventory y, planning horizon n

and optimal inventory management

the recursion applies (with given boundary condition $v_0(y) = v_0 = 0$)

a) in the BACKORDER case: n = 1, 2, 3, ...

$$v_n(y) = \min_{x \geq y} \{ k\delta(x - y) + a(x - y) + f(x) + \rho \int_0^\infty v_{n-1}(x - u)\,dP(u) \}$$

(36.2)

b) in the LOST SALES case: n = 1, 2, 3, ...

$$v_n(y) = \min_{x \geq y} \{ k\delta(x - y) + a(x - y) + f(x) + \rho \int_0^x v_{n-1}(x - u)\,dP(u) +$$

$$+ [1 - P(x)]v_{n-1}(0) \} \quad .$$

(36.3)

The Theory of Dynamic Programming (RICHARD BELLMAN) was first used to develop the AHM–Model. Hence the above recursion method was called the BELLMAN Principle of Optimality and, both equations (36.2) and (36.3) the BELLMAN Functional Equations.

The stock level immediately before a decision is given by y. One can also formulate the functional equations related to a state z   : Inventory after a decision (BECKMANN (1968)):

$$v_n(z) = f(z) + \rho \int_0^\infty \underset{x-y}{\text{Min}} \{k\delta(x-y) + a(x-y)$$
$$+ v_{n-1}(z+x-y-u)\}dP(u) . \qquad (36.4)$$

This form is advantageous for numerical purposes since the state space is smaller. However, it is not pursued here.

§37 THE AHM–MODEL IN THE STATIONARY CASE

The AHM Model in the stationary case is more amenable to an analytical treatment. Aside from this, the functional equations are simplified because the iteration index n is disregarded.

Several times previously, we used the method for the stationary case which takes out the proportional ordering cost component of the cost function to be minimized. In the long run, the receipt of goods must be equal to the withdrawal of goods, i.e., the expected stationary mean order quantity in the BACKORDER case is equal to the expected value μ of demand in one period and hence independent of the ordering rule.

We now want to use the same consideration for the stationary AHM Model. The value function (BACKORDER Case) is

$$v(y) = \underset{x \geq y}{\text{Min}} \{k\delta(x-y) + a(x-y) + f(x) + \rho \int_0^\infty v(x-u)dP(u)\} . \qquad (37.1)$$

In order to consider all costs at their present value these are discounted by $\rho < 1$.

The expected value of all proportional costs are now to be removed from v. As an approximation, it is assumed that at the start of each period the expected sales μ of the previous period is ordered. Then the expected value of all proportional ordering costs is

$$a\mu \frac{\rho}{1 - \rho} - ay \ . \tag{37.2}$$

We substitute the adjusted value function $\hat{v}(y)$

$$\hat{v}(y) := v(y) - a(\frac{\mu\rho}{1 - \rho} - y) \tag{37.3}$$

in the functional equation (37.1)

$$\hat{v}(y) + a(\frac{\mu\rho}{1 - \rho} - y) = \underset{x \geq y}{\text{Min}} \{k\delta(x - y) + a(x - y) + f(x) + \rho \int_0^\infty [\hat{v}(x - u)$$

$$+ a(\frac{\mu\rho}{1 - \rho} - x + u)] dP(u)\}$$

$$= \underset{x \geq y}{\text{Min}} \{k\delta(x - y) + \underbrace{ax - \rho ax + f(x) + a\mu \frac{\rho^2}{1 - \rho} + \rho a\mu}_{=: \hat{f}(x)} \underbrace{- ay +}_{= a(\frac{\mu\rho}{1 - \rho} - y)}$$

$$+ \rho \int_0^\infty \hat{v}(x - u) dP(u)\}$$

and obtain (in the BACKODER case)

$$\boxed{\hat{v}(y) = \underset{x \geq y}{\text{Min}} \{k\delta(x - y) + \hat{f}(x) + \rho \int_0^\infty \hat{v}(x - u) dP(u)\}} \tag{37.4}$$

with

$$\hat{f}(x) = ax(1 - \rho) + f(x) \ . \tag{37.5}$$

In the LOST SALES case, this trick cannot be applied since the average order quantity per period is less than the expected demand.

In the following sections we will deal with two questions:

1. How can one obtain a concrete solution for a single case?
2. How does the structure of the optimal solution look like if one can, at all, speak of a structure?

§38 STANDARDIZATION

The AHM Model can be standardized under certain assumptions.

Condition (C1):
The demand u has a probability distribution of the form $P(u; \mu, \sigma) = Q(\frac{u - \mu}{\sigma})$ which can be standardized.

We then set

$$u = \mu + \sigma \epsilon \tag{38.1}$$

μ : Expected value of demand
σ : Standard deviation of u
ϵ : Random variable with $\mu_\epsilon = 0$, $\sigma_\epsilon = 1$ (stochastic component of demand)

$$q(\frac{u - \mu}{\sigma}) = p(\epsilon).$$

Condition (C2):
Let the fixed order cost k be of the form

$$k = k_0 \sigma . \tag{38.2}$$

The Principle of Optimality can be formulated in the following manner

$$v(y) = \underset{x \geq y}{\text{Min}} \{ k\delta(x-y) + a(x-y) + h \int_{u=0}^{x} (x-u)q(\tfrac{u-\mu}{\sigma})d(\tfrac{u}{\sigma}) +$$

$$+ g \int_{x}^{\infty} (u-x)q(\tfrac{u-\mu}{\sigma})d(\tfrac{u}{\sigma}) \} + \rho \int v(x-u)q(\tfrac{u-\mu}{\sigma})d(\tfrac{u}{\sigma}) . \tag{38.3}$$

The right–hand side becomes proportional to σ under condition (C2). To show this, we perform the following variable transformations

$$\xi := \tfrac{x-\mu}{\sigma} ; \tag{38.4}$$

$$\eta := \tfrac{y}{\sigma} . \tag{38.5}$$

In addition, $d\epsilon = d(\tfrac{\mu}{\sigma})$ and we define

$$\sigma\nu(\eta) := v(y) . \tag{38.6}$$

Then from (38.3)

$$\sigma\nu(\eta) = \underset{\xi \geq \eta}{\text{Min}} \{ k_0\sigma\delta(\xi-\eta) + a\sigma(\xi-\eta) + \sigma f(\xi) +$$

$$+ \rho \int \sigma\nu(\xi-\epsilon)p(\epsilon)\,d\epsilon \} . \tag{38.7}$$

The factor σ is cancelled and one obtains the STANDARDIZED EQUATION

$$\boxed{\begin{aligned} \nu(\eta) &= \underset{\xi > \eta}{\text{Min}} \{ k_0\delta(\xi-\eta) + a(\xi-\eta) + f(\xi) + \\ &\quad + \rho \int \nu(\xi-\epsilon)p(\epsilon)d\epsilon \} \end{aligned}} \tag{38.8}$$

The derivation shows:

1.   The expected inventory and shortage costs are proportional to the standard deviation of demand and independent of its expected value. v increases with increasing σ.

2.   Once and for all the standardized problem for different cost rates $\frac{h}{g}$ and $\frac{k_o}{g}$ is solved. With the help of the reverse transformation

$$x = \mu + \sigma \xi ,$$

the optimal policy of the given problem is derived from the optimal policy s*, S* and is given by:

$$\boxed{\begin{aligned} s &= \mu + \sigma s^* \; ; \\ S &= \mu + \sigma S^* \, . \end{aligned}}$$

The conditions (C1) and (C2), however, must be checked.

§39   EXPONENTIALLY DISTRIBUTED DEMAND

We attempt to hold on to our previous solution schema:

1.   The structure of the optimal ordering rule in parametric form is given;
2.   Derivation of the cost function c;
3.   Minimization of v with respect to the parameter of the ordering rule determines the optimal ordering rule.

The objective will only be achieved if two conditions are fulfilled:

—   stationary model,
—   the optimal ordering rule has the assumed parametric structure.

Again, we first postulate the structure of the ordering rule and optimize only within this structure. (It will be shown later that the global optimal ordering rule has the assumed structure.)

Assumption: Let the ordering rule be an (s, S) policy

$$\boxed{\begin{array}{ll} s < y \leq S: \text{do nothing} & ; \\[2mm] y \leq s: & \text{replenish up to S.} \end{array}}$$

(39.1)

If the initial stock y > S, one waits until the stock has fallen down to S. The (s, S) policy then applies at that point and S is the maximum stock level. The value function, restricted to the class of the (s, S) policy, is (in the BACKORDER Case)

$$\hat{v}(y) = \begin{cases} \hat{f}(y) + \rho \int_0^\infty \hat{v}(y - u)\,dP(u) , & \text{for } y \geq s ; \\[4mm] k + \hat{f}(S) + \rho \int_0^\infty \hat{v}(S - u)\,dP(u) , & \text{for } y \leq s . \end{cases}$$

(39.2)

(39.3)

The proportional ordering costs are ignored in this model. It has been shown in §37 that this is not a relevant restriction. We assume a continuous state space such that the two alternatives ("order" and "do not order") are both acceptable. The two equations (39.2) and (39.3) hold for y = s.

For y ≤ s, the value function (39.3) is independent of y. It applies specifically to

$$\hat{v}(y) = \hat{v}(s) , \qquad y \leq s.$$

Hence, (39.2) can also be written as

$$\boxed{\hat{v}(y) = \hat{f}(y) + \rho \int_0^{y-s} \hat{v}(y-u)\,dP(u) + \rho\hat{v}(s)\,[1 - P(y-s)] , \quad y \geq s .}$$

(39.4)

For y = s we have $\hat{v}(s) = \hat{f}(s) + \rho\hat{v}(s)$ or

$$\hat{v}(s) = \frac{\hat{f}(s)}{1-\rho} .$$

(39.5)

In general, the preceding considerations still apply. We now attempt to solve the integral equation (39.4) for an exponentially distributed demand.
We have,

$$P(u) = 1 - e^{-\alpha u} ,$$

$$p(u)du = \alpha e^{-\alpha u} du ,$$

$$E\{u\} = \mu = \frac{1}{\alpha} .$$

The expected inventory and shortage costs are

$$\hat{f}(x) = f(x) + ax(1 - \rho)$$

$$f(x) = (h + g)[x - \frac{1}{\alpha}(1 - e^{-\alpha x})] + g(\frac{1}{\alpha} - x) . \tag{39.6}$$

We substitute it in (39.4), perform a variable transformation $\xi = y - u$ in the integral and multiply the equation with $e^{\alpha y}$. This gives

$$\hat{v}(y)e^{\alpha y} = (h + g)[ye^{\alpha y} - \frac{1}{\alpha}(e^{\alpha y} - 1)] + \frac{g}{\alpha}e^{\alpha y} + (a - g - \rho)ye^{\alpha y} +$$

$$+ \alpha\rho \int\limits_{s}^{y} \hat{v}(\xi)e^{\alpha\xi} d\xi + \alpha\rho\hat{v}(s)e^{\alpha s} . \tag{39.7}$$

Using the definition

$$w(y) := \hat{v}(y)e^{\alpha y},$$

we have

$$w(y) = (h + g)[ye^{\alpha y} - \frac{1}{\alpha}(e^{\alpha y} - 1)] + \frac{g}{\alpha}e^{\alpha y} + (a - g - \rho)ye^{\alpha y} +$$

$$+ \alpha\rho \int\limits_{s}^{y} w(\xi)d\xi + \alpha\rho w(s) .$$

By differentiation we convert this integral equation in a differential equation of the following form

$$w'(y) - \alpha\rho w(y) = [\alpha y(h + a - \rho) + a - \rho]e^{\alpha y} .$$

With the integrating factor $e^{-\alpha\rho y}$, the differential equation becomes

$$\frac{d}{dy}(w(y)e^{-\alpha\rho y}) = [\alpha y(h + a - \rho) + (a - \rho)]e^{\alpha(1 - \rho)y} .\tag{39.8}$$

We integrate (39.8) and again obtain the original cost function $\hat{v}$ by multiplying both sides by $\exp(\alpha\rho y - \alpha y)$. In the second step we substitute the boundary condition (39.5) and get the required solution for $\hat{v}$.

In the third step, $\hat{v}$ is to be minimized with respect to $y = S$ and s

$$\underset{y}{\text{Min}}\ \hat{v}(y) = \hat{v}(S) \qquad\qquad \Rightarrow\ S$$

$$\underset{s}{\text{Min}}\ \hat{v}(y) \qquad\qquad \Rightarrow\ s .$$

However, the conditions $\dfrac{d\hat{v}}{dy} = 0$ and $\dfrac{d\hat{v}}{ds} = 0$ necessary for the minimum lead to transcendental equations so that no explicit solutions can be given for s and S.

To arrive at a solution we modify the cost structure. Let

$$f(x) = hx + g \int_{x}^{\infty} (\xi - x)e^{-\xi}\,d\xi = hx + ge^{-x} .\tag{39.9}$$

In addition, let the expected value of demand $\mu = 1$ which can be obtained by rescaling the units of demand without limitations. Then $\alpha = 1$. Inventory cost is taken in relation to the start of the period and shortage cost at the end. Inventory $y > 0$ is therefore given a higher cost than before. Instead of (39.7) we now obtain

$$\hat{v}(y) = hy + ge^{-y} + \frac{\rho}{1-\rho}(hse^{s} + g)e^{-y} + \rho \int_{s}^{y} \hat{v}(\xi)e^{\xi-y} d\xi .$$ (39.10)

The solution of this equation is

$$\hat{v}(y) = \frac{hy}{1-\rho} - \frac{\rho h}{(1-\rho)^2} + [\frac{\rho h}{(1-\rho)^2} + \frac{ge^{-s}}{1-\rho}] e^{(\rho-1)(y-s)} .$$ (39.11)

$\hat{v}$ is convex in y and s. $S = y$ is the optimal initial stock. Hence, the minimum of $\hat{v}$ lies at S

$$\hat{v}'(S) = 0.$$ (39.12)

If one substitutes $y = S$ and $y = s$ in (39.2) and (39.3), respectively, and subtracts both expressions one gets

$$\hat{v}(s) - \hat{v}(S) = k.$$ (39.13)

We use both these equations (39.12) and (39.13) to calculate s , S, and $D = S - s$. For $y = s$

$$\hat{v}'(S) = \frac{h}{1-\rho} + (\rho - 1)[\frac{\rho h}{(1-\rho)^2} + \frac{ge^{-s}}{1-\rho}] e^{(\rho-1)(S-s)} .$$

Since $v'(S) = 0$, it follows that

$$D = S - s = \frac{1}{1-\rho} \ln[\rho + (1-\rho) \frac{g}{h} e^{-s}] .$$ (39.14)

From (39.13) we have

$$\frac{\rho h}{(1-\rho)^2} + \frac{ge^{-s}}{1-\rho} - \frac{hD}{1-\rho} - [\frac{\rho h}{(1-\rho)^2} + \frac{ge^{-s}}{1-\rho}] e^{(\rho-1)D} = k$$ (39.15)

or

$$[\rho + (1 - \rho)\tfrac{g}{h} e^{-s}] - (1 - \rho)D - [\frac{(1 - \rho)g}{h} e^{-s} + \rho]e^{(\rho-1)D}$$

$$= \frac{(1 - \rho)^2}{h} k \ . \tag{39.16}$$

Using the shortened form

$$q := \rho + (1 - \rho)\tfrac{g}{h} e^{-s}$$

(39.14) becomes

$$D = \frac{1}{1 - \rho} \ln q$$

and (39.16) takes the form

$$q = 1 + \frac{(1 - \rho)^2}{h} k + \ln q \ . \tag{39.17}$$

The values s and D are computed as follows: at first, q is determined iteratively from equation (39.17). Then s, S and D are given by

$$s = \ln \frac{g \cdot (1 - \rho)}{h \cdot (q - \rho)} \ ; \tag{39.18}$$

$$D = \frac{1}{1 - \rho} \cdot \ln q \ ; \tag{39.19}$$

$$S = s + D. \tag{39.20}$$

§40 OPTIMALITY OF THE (s ,S)– POLICY

The optimality of a (s, S) policy (S = s + D) was obvious for models with continuous stock review. Because of this, the analytical solution method was considerably simplified. Does it also apply for the AHM–Model?

We consider the AHM Model in the backorder case. The Principle of Optimality is given by

$$v_n(y) = \underset{x \geq y}{\text{Min}} \; \{k\delta(x-y) + a(x-y) + f(x) + \rho \int_0^\infty v_{n-1}(x-u)dP(u)\}, \qquad (40.1)$$

$$u = 1, 2 \ldots, N.$$

By evaluating this recursion, a solution can always be found. One begins with a starting value v_0 defined from the problem and computes in sequence the chain

$$v_0 \Rightarrow v_1 \Rightarrow v_2 \Rightarrow \ldots \Rightarrow v_n \Rightarrow \ldots \Rightarrow v_N \; .$$

One calls this the VALUE ITERATION of dynamic programming. A separate optimal ordering rule is obtained for each period. For an infinite planning horizon a suitable termination criterion is still to be defined. (For a stationary model, however, one switches to another method.) Value iteration requires a great deal of numerical computation. For example, if y varies between –1000 and 1000, then the minimization operation must be performed 2001 times for each n. For a single minimization step, the right–hand side of (40.1) is evaluated 1000 times on the average.

The optimal ordering rule can considerably be simplified for a given structure. For a (s_n, S_n) structure, for example, one first takes the minimum around y = –1000. One knows that an order will certainly be placed. This minimization step gives s_n.

For all remaining stocks y = −999, −998, ... minimization is reduced to a comparison between the two alternatives "do not order" and "fill inventory up to S_n". As soon as s_n is known, minimization is dropped for the remaining y values y = s_{n+1}, s_{n+2}, ..., S_n.

With this viewpoint, the question "When is a (s_n, S_n) policy optimal?" gains importance. It should be examined in the following manner:

We rewrite (40.1) in order to formulate the problem more transparently. We remove the cost −ay from the minimization and obtain

$$v_n(y) = -ay + \underset{x \geq y}{\text{Min}} \underbrace{\{k\delta(x-y) + ax + f(x) + \rho \int v_{n-1}(x - u)dP(u)\}}. \qquad (40.2)$$

$$=: H_n(x)$$

We now split the functional equation into two alternatives, (I) and (II)

$$v_n(y) = -ay + \begin{cases} H_n(y) , & \text{if no order (I);} \\ k + \underset{x \geq y}{\text{Min }} H_n(x) , & \text{if order (II);} \end{cases} \qquad (40.3)$$

and obtain the decision rule:

$$\text{in case, } H_n(y) - \underset{x \geq y}{\text{Min }} H_n(x) =: \Delta H_n \begin{cases} < k \Rightarrow \text{ do nothing ;} \\ > k \Rightarrow \text{ order x-y;} \end{cases} . \qquad (40.4)$$

For ΔH_n = k, both alternatives are equally acceptable and we may choose either one.

A physical interpretation is offered in the following manner. For the inventory y to be affected by external intervention (order), it is necessary to overcome the friction by exerting a certain amount of force k. This would only pay off if k is less than the amount of potential energy ΔH to be released.

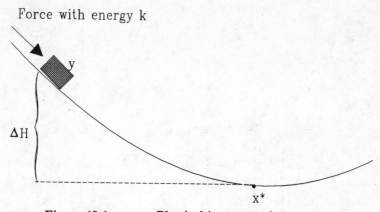

Figure 40.1:   Physical interpretation:

To overcome its friction, the point mass y experiences a momentum k and slides to x*. Hence, a potential energy ΔH is released.

The quantity y which makes it worthwhile to order depends on the form of H_n. Two examples are shown in Fig. 40.2 and 40.3. Each of the order areas are denoted with hatched lines. The global minimum of H_n at S_n determines – without considering a possible higher initial stock y – the maximum inventory.

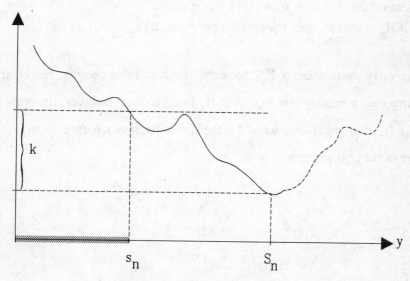

Figure 40.2: (s_n, S_n) policy

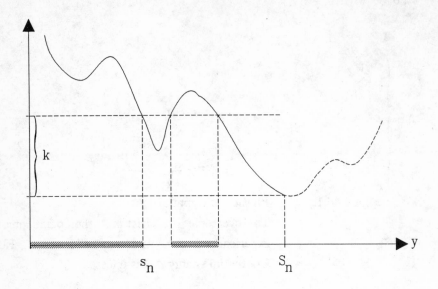

Figure 40.3: A more complicated policy

This consideration makes two things clear:

1.   If $k = 0$, one increases the inventory to the optimal value at each stock review even if the deviation from this value is small.

2.   If H_n is convex, then the optimal policy has a (s_n, S_n) structure.

The convexity assumption of H_n, however, is not suitable because, firstly, it is more restrictive than necessary (see Fig. 40.2; H_n is not convex; however, the optimal policy is of the (s_n, S_n) type) and, secondly, the convexity is not passed on to H_{n+1}. One sees this in the next example.

For $n = 1$,

$$v_1(y) = -ay + \begin{cases} H_1(y) & \text{(I)} \\ k + H_1(S_1) & \text{(II)} \end{cases}$$

If f(x) is convex $\Rightarrow H_1 = ay + f(y)$ is also convex.

For example, the following graph of v_1 is given below

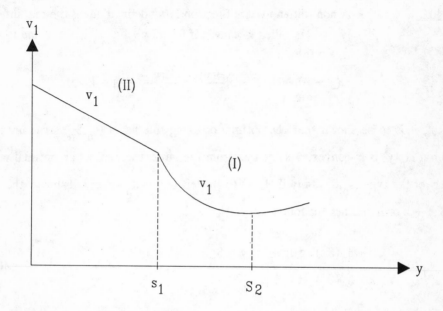

Figure 40.4: Example for a non–convex but a k–convex v_1

$v_1^{(I)}$ is convex, $v_1^{(II)}$ is a linear function with slope –a. For some values of a, v_1 is not convex for all y. Hence,

$$H_2(y) = ay + f(y) + \rho \int v_1(y-u)\,dP(u)$$

is also no longer convex for all y.

However, one achieves the objective if the concept of convexity is generalized in a suitable way.

Def. 40.1:   A function $H(y)$ defined along the real interval $[y^-;y^+]$ is called
                    k–convex if for all $y \, \epsilon \, [y^-;y^+]$ and for any $\alpha, k \geq 0$:

$$H(y + \alpha) - H(y) - \alpha H'(y) + k \geq 0.$$

Def. 40.2:   A non–differentiable function $H(y)$ defined along the real interval
                    $[y^-;y^+]$ is called k–convex if for all $y \, \epsilon \, [y^-;y^+]$, $\beta > 0$ and for any
                    $\alpha, k \geq 0$:

$$H(y + \alpha) - H(y) + \alpha[\frac{H(y) - H(y - \beta)}{\beta}] + k \geq 0.$$

At first, it is to be shown that the optimal ordering rule has a (s_n, S_n) structure for a
fixed n if $H_n(y)$ is k–convex. As one sees from Fig. 40.2, there exists an optimal policy
exactly of the type (s_n, S_n) only if $H_n(y)$ to the left of $y = s_n$ never falls below the level
$H_n(S_n) + k$ or it reaches the following:

$$H_n(y) \begin{cases} < k + H_n(S_n), \text{ for } s_n < y < S_n \; ; \\ > k + H_n(S_n), \text{ for } y < s_n \quad . \end{cases} \tag{40.5}$$

This corresponds exactly to the decision rule (40.4). The condition (40.5) is satisfied
with certainty if $H_n(y)$ falls monotonically for $y \leq s_n$, i.e., if

$$H'_n(y) < 0 \text{ for } y \leq s_n. \tag{40.6}$$

The k–convex functions $H_n(y)$ satisfy this requirement (40.6). We shall prove this by
contradiction. We assume the opposite:

The k–convex function $H_n(y)$ has to the left of s_n a relative maximum $H_n(y_1)$, $y_1 <$
s_n, $H_n(y_1) > k + H_n(S_n)$ (see the following Figure (40.5))

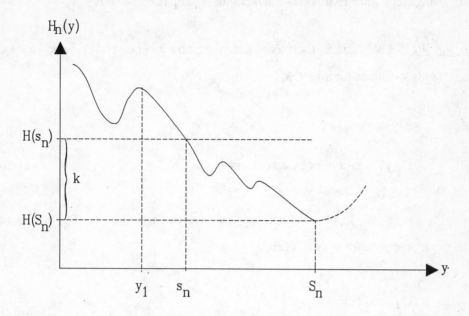

Figure 40.5: Proof of the optimality of the (s_n, S_n) policy if H_n is k–convex

Let s_n be the largest y value for which $H_n(y)$ exceeds the level $H(s_n) = H(s_n)+k$ (in this case, $H'_n(s_n) < 0$). Because of the relative maximum at y_1, H_n cannot be k–convex since the defining inequality is not satisfied. If we choose $S_n = y_1 + \alpha$, then it follows that

$$\underbrace{H_n(S_n) - H_n(y_1)}_{< -k} - \underbrace{\alpha H'_n(y_1)}_{= 0} + k < 0 \ .$$

This is a contradiction to the condition of k–convexity of H_n. Hence, for $y < s_n$, a k–convex function H_n can not have a relative extremum. Since $H'_n(s_n) < 0$, $H_n(y)$ is therefore monotonically decreasing for all $y \le s_n$, i.e., the (s_n, S_n) policy is optimal.

The following steps show that the k–convexity of H_n is passed on to H_{n+1}.

<u>Statement 40.1:</u> The basis is the model described above. Let $H_1(y)$ be k–convex. Then $H_n(y)$ is k–convex for all $n \in \mathbb{N}$.

<u>Proof:</u>

n = 1: $H_1(y)$ is k–convex by assumption.

n > 1:   Let $H_n(y)$ be k–convex.

$\Rightarrow$ (s_n, S_n) policy is optimal, i.e., we can use the (s_n, S_n) policy to compute for $v_n(y)$. Then

$$(*) \quad v_n(y) = -ay + \begin{cases} H_n(y), & \text{for } y > s_n ; \\ k + H_n(S_n), & \text{for } y \le s_n . \end{cases}$$

We compute $v_n(y)$.

1. $\underline{s_n < y \le S_n}$: $v_n(y) = -ay + H_n(y)$ is k–convex.

2. $\underline{y \le s_n < y + \alpha \le S_n}$: It must be shown that

$$v_n(y + \alpha) - v_n(y) - \alpha v_n'(y) + k \ge 0 .$$

Substituting (*) gives

$$-a(y + \alpha) + H_n(y + \alpha) + ay - k - H_n(S_n) + \alpha a + k \ge 0$$
$$H_n(y + \alpha) - H_n(S_n) \ge 0.$$

This is always correct since $H_n(S_n) = \underset{y}{\text{Min }} H_n(y)$.

3. $y < y + \alpha < s_n$: $v_n(y) = -ay + k + H_n(S_n)$ is linear,

therefore, also k–convex.

$\Rightarrow$ $v_n(y)$ is k–convex

$\Rightarrow$ $\int_0^\infty v_n(x - u)dP(u)$ is k–convex

$\Rightarrow$ $H_{n+1}(y) = -ay + f(x) + \int v_n(x-u)dP(u)$ is k–convex, since ay and f(x) are convex.

Since $H_1(x) = ax + f(x)$ is k–convex, it was therefore shown that

> The AHM Model (BACKORDER Case) developed in §36 has an optimal ordering rule of a (s_n, S_n) policy type
>
> for <u>each</u> period.

The important point here is the form of the expected inventory and shortage cost. Again it is given as

$$f(y) = h \int_0^y (y - u)dP(u) + g \int_y^\infty (u - y)dP(u) .$$

The inventory level as well as the shortage quantities are covered by proportional costs h and g, respectively. Originally, ARROW, HARRIS, MARSCHAK, KARLIN, SCARF, BECKMANN and others worked with shortage costs which were equally high (tailored to the situation in the Navy, compare §26.2). Inevitably, all efforts to arrive at optimal (s_n, S_n) policies failed.

Later, numerous generalizations of the above function f(y) were given, but the (s_n, S_n) policies still remained among them. (e.g., VEINOTT (1966), SCHÄL (1976)).

§41   ELIMINATION OF PROPORTIONAL ORDERING COSTS
WITH FINITE PLANNING HORIZON

We attempt to integrate the proportional ordering cost $a(x - y)$ into the inventory and shortage costs. That would be possible immediately if the proportional ordering cost was a linear function of the inventory x as in the case with the inventory and shortage costs, hx and $-gx$, respectively. To achieve this, we choose the approach:

Formula: $\hat{v}_n(y) = v_n(y) + ay - a\mu$, $\mu = E\{u\}$.

The proportional ordering cost for $\hat{v}_n(y)$ is now

$$a(x - y) + ay - a\mu. \tag{41.1}$$

Idea: The term ay is cancelled and ax remains. $a\mu$ is a constant correction term. With this approach we formulate the Principle of Optimality. The functions $v_n(y)$ follow that of equation (36.2). We have

$$\hat{v}_n(y) = \min_{x \geq y} \{k\delta(x - y) + ax - ay + f(x) +$$

$$+ \rho \int [\hat{v}_{n-1}(x - u) - a(x - u) + a\mu]\,dP(u)\} + ay - a\mu$$

$$= \min_{x \geq y} \{k\delta(x - y) + a(1 - \rho)(x - \mu) + f(x) + \rho \int \hat{v}_{n-1}(x - u)\,dP(u)\} + \rho a\mu$$

A term $\rho a\mu$ still remains. To be able to eliminate this term we extend this approach to include a variable correction term,

extended formulation: $\hat{v}_n(y) = v_n(y) + a(y - \mu) + b_n$, $\qquad(41.2)$

and return to the functional equation (36.2):

$$\hat{v}_n(y) = \underset{x \geq y}{\text{Min}} \, \{k\delta(x - y) + a(x - \mu) + b_n + f(x)$$

$$+ \rho \int [\hat{v}_{n-1}(x - u) - a(x - u) + a\mu - b_{n-1}] \, dP(u)\}$$

$$= \text{Min} \, \{k\delta(x - y) + a(1 - \rho)(x - \mu) + f(x)$$

$$+ \rho \int \hat{v}_{n-1}(x - u) \, dP(u)\} + \underbrace{b_n + \rho a\mu - \rho b_{n-1}}_{\overset{!}{=} \, 0} \, .$$

The remaining term $b_n + \rho a\mu - \rho b_{n-1}$ must be eliminated. Hence it must follow that

$$b_n = \rho(b_{n-1} - a\mu)$$

$$= \rho[\rho b_{n-2} - a\mu) - a\mu]$$

$$\vdots$$

$$= \rho^n b_0 - \rho a\mu \, \frac{1 - \rho^n}{1 - \rho} \, .$$

We choose as initial value $b_0 = 0$ and obtain

$$\boxed{b_n = -\rho a\mu \, \frac{1 - \rho^n}{1 - \rho} \, .} \qquad\qquad (41.3)$$

Hence the remaining term is eliminated.

In the last step we integrate the order cost in the cost function $f(x)$. Let the new function be $\hat{f}(x)$

$$\hat{f}(x) := a(1 - \rho)(x - \mu) + f(x) \, . \qquad\qquad (41.4)$$

Previously, it was

$$f(x) = (h + g) \int_0^x P(u) du + g(\mu - x) \ .$$

Hence,

$$\hat{f}(x) = (h + g) \int_0^x P(u) du + \underbrace{[g - a(1 - \rho)]}_{=: \; \hat{g}} (\mu - x) \ . \tag{41.5}$$

We define

$$\hat{h} := h + a(1 - \rho) \ ; \tag{41.6}$$

$$\hat{g} := g - a(1 - \rho) \tag{41.7}$$

as the new cost rates for the inventory and shortage quantities, repectively, and obtain
the results

$$\boxed{f_{\hat{g}, \hat{h}}(x) = \hat{f}(x)} \tag{41.8}$$

$$\boxed{\hat{v}_n(y) = v_n(y) + a(y - \mu) - \rho a \mu \, \frac{1 - \rho^n}{1 - \rho}} \tag{41.9}$$

$$\boxed{\hat{v}_n(y) = \underset{x \geq y}{\text{Min}} \; \{k\delta(x - y) + \hat{f}(x) + \rho \int_0^\infty \hat{v}_{n-1}(x - u) dP(u)\} \ .} \tag{41.10}$$

Comparison with the elimination of proportional ordering cost
in the stationary model (§37)

The above method can be extended to the case $n = \infty$.

For $n \to \infty$, (41.3) becomes

$$\lim_{n \to \infty} b_n =: b = \frac{-\rho a \mu}{1 - \rho} , \tag{41.11}$$

and from (41.2)

$$\begin{aligned} \hat{v}(y) &:= v(y) + a(y - \mu) + b \\ &= v(y) + ay - \frac{a\mu}{1 - \rho} . \end{aligned} \tag{41.12}$$

For the stationary model in §(37) we used a transformation

$$\hat{v}(y) := v(y) + ay - \frac{\rho a \mu}{1 - \rho} .$$

There it was

$$\hat{f}(x) := f(x) + a(1 - \rho)x .$$

The difference between the two formulations is that in the stationary model the average ordering cost $a\mu$ is always taken <u>at the end</u> of the period so that it can be discounted. Therefore, the constant correction term is

$$\frac{\rho a \mu}{1 - \rho} .$$

The formulation chosen in this section uses the ordering cost <u>at the start</u> of the period. Hence the correction term here is

$$\frac{a\mu}{1 - \rho} .$$

Finally, the difference in the total cost between both models is that in the latter model, the average ordering cost $a\mu$ occurs more often, namely to stock up for the first period. This is also shown by the following formal consideration.

A comparison of both transformations shows

$$\Delta v := \overset{\approx}{v}(y) - \hat{v}(y) = a\mu \; ;$$

$$\Delta f := \overset{\approx}{f}(y) - \hat{f}(y) = a\mu(1 - \rho) \; .$$

The difference Δv of the value functions can be explained completely by the different single period costs. It is, namely, the present value of the differences Δf for all periods

$$\Delta f \sum_{n=0}^{\infty} \rho^n = a\mu$$

which is exactly the difference Δv.

The advantage of the formulation chosen in this section is that the elimination of the proportional ordering cost can be interpreted as a modification of the inventory and shortage cost rates h and g.

§42   BOUNDS FOR (s_n, S_n)

Bounds for the (s_n, S_n) policies will be derived in this section.

$$\underline{s} \leq s \leq \overline{s} \; < \; \underline{S} \leq S \leq \overline{S} \; . \tag{42.1}$$

This is useful for numerical purposes. One can limit oneself to x values in a minimum search $\underset{x \geq y}{\text{Min}} \{ \}$ which lie within the intervals $[\underline{s}, \overline{s}]$ and $[\underline{S}, \overline{S}]$, respectively. Besides, the bounds are used for the proof of an optimal (s, S) policy for the stationary model. More of this later.

1. Bound $\underline{S}$:

The replenishment point S_1 of the single period model will be a lower limit S for the replenishment point S_n of all multiple period models, n > 1.

A plausible justification for this is as follows:

Since the fixed cost k > 0, in a multiple period model one will stock up for the future at the start of the first period. In a single period model stocking up for succeeding periods is not necessary. A rigorous proof will be given to show that this assumption is correct. It is already shown that for all periods a (s_n, S_n) policy is optimal, i.e.,

$$
v_n(y) = -ay \begin{cases} + H_n(y) & , \quad \text{for } y > s_n \ ; \\ + k + H_n(S_n) & , \quad \text{for } y \le s_n \ . \end{cases}
\tag{42.2}
$$

We derive the proof of the model with proportional cost. $v_n(y)$ is differentiable except at the point $y = s_n$.

For n = 1

$$
H_1(y) = \underbrace{ay + f(y)}_{=:\, G(y)}
\tag{42.3}
$$

G(y) is the single period cost. It is independent of n. Take H_1 as its minimum at point S_1, i.e.,

$$
G'(y) < 0 \quad \text{for all } y < S_1.
\tag{42.4}
$$

We now show by induction: $H_n'(y) < 0$ for all $y < S_1$, $n \in \mathbb{N}$.

$H_1'(y) < 0$ for all $y < S_1$ was already shown.

Let $H_n'(y) < 0$ for all $y < S_1$. Then from (42.2)

$$v_n'(y) = \left\{ \begin{array}{c} -a + H_n'(y) \\ -a \end{array} \right\} \leq -a < 0 \text{ for all } y < S_1.$$

With

$$H_{n+1}'(y) = G'(y) + \rho \int v_n'(x - u) dP(u)$$

one obtains

$$H_{n+1}'(y) \leq G'(y) - \rho a < 0 \text{ for all } y < S_1. \tag{42.5}$$

Hence $H_n(y)$ assumes its minimum at $y \geq S_1$ for all $n \in \mathbb{N}$. The minimizing value of H_n is the replenishment point S_n in the nth period counting backwards (i.e., the first period of an n–period model). It is therefore

$$S_1 \leq S_n \text{ for all } n \in \mathbb{N}.$$

(Bound due to IGLEHART (1963))

Computation of S_1:

$$H_1(S_1) = \underset{x}{Min}\, H_1(x) \iff ax + f(x) \to \underset{x}{Min}$$

$$\iff ax + (h + g) \int_0^x P(u)\,du + g(\mu - x) \to \underset{x}{Min}$$

$$\iff a + (h + g)P(x) - g = 0$$

$$S_1 = P^{-1}\left(\frac{g - a}{g + h}\right) .$$

The bound $\underline{S}$, however, can still be improved if one uses the model with the proportional ordering cost eliminated. We have

$$\hat{H}_1(y) = \hat{f}(y)$$

$$\hat{v}_n'(y) = \left\{ \begin{matrix} \hat{H}_n'(y) \\ 0 \end{matrix} \right\} \le 0 \text{ for all } y < \hat{S}_1 .$$

$\hat{S}_1$ is the minimizer of $\hat{H}_1(y) = \hat{f}(y)$. Following the schema of the proof given above one obtains

$$\hat{H}_n(y) \le \hat{f}'(y) < 0 \text{ for all } y < \hat{S}_1 .$$

Therefore $\hat{S}_1$ is also a lower bound for S_n

$$\hat{S}_1 \le S_n , \qquad \text{for all } n \in \mathbb{N}.$$

(Bound due to VEINOTT.)

$\hat{S}_1$, however, is stricter than S_1 as shown by the following computation:

$$\hat{H}_1(\hat{S}_1) = \underset{x}{\text{Min}} \ \hat{H}_1(x) \iff \hat{f}(x) \to \underset{x}{\text{Min}}$$

$$\iff f(x) + a(1-\rho)(x-\mu) \to \underset{x}{\text{Min}}$$

$$\iff (h+g)P(x) - g + a(1-\rho) = 0$$

$$\boxed{\underline{S} = \hat{S}_1 = P^{-1}\left(\frac{g - a(1-\rho)}{g+h}\right)} \ . \tag{42.6}$$

Since the distribution function $P(u)$ of demand always increases monotonically, $\hat{S}_1 \geq S_1$. The advantage of eliminating the proportional ordering cost as proposed by BECKMANN is evident here (compare §36).

This fact is graphically represented in Fig. 42.1. The function $\hat{f}$ has its minimum farther to the right than f, but falls flatter to the left than f.
(Reason: $\hat{g} = g - a(1-\rho) < g$, i.e., the shortage cost is lower.)

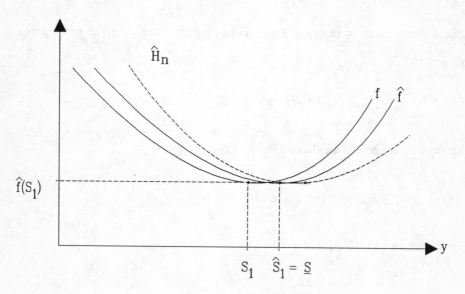

Figure 42.1: Lower bound $\underline{S}$

$\hat{H}_n$ falls to the left of $\hat{S}_1$ steeper than $\hat{f}$.

Since it only depends on the height differences ΔH and Δf, respectively, $\hat{H}_n$ and f are moved in the above figure such that its minimum lies at equal height with $\hat{f}(\hat{S}_1)$. Then $\hat{H}_n$ lies over $\hat{f}$ for all $y < \hat{S}_1$. Also, H_n lies over f for all $y < S_1$.

2. Boundary $\bar{s}$:

We write the Principle of Optimality in the form

$$\hat{v}_n(y) = \text{Min}\ \{\hat{H}_n(y)\ k + \underset{x \geq y}{\text{Min}}\ \hat{H}_n(x)\}$$

and first show that

$$\hat{v}_n(y) \leq \hat{v}_n(y') + k\ ,\quad \text{for } y \leq y'\ . \tag{42.7}$$

We have

$$\hat{v}_n(y) \leq k + \underset{x \geq y}{\text{Min}}\ \hat{H}_n(x) \leq k + \underset{x \geq y'}{\text{Min}}\ \hat{H}_n(y') \leq k + \hat{v}_n(y')\ \text{for } y \leq y'.$$

With the help of this inequality we obtain

$$\hat{H}_n(y) - \hat{H}_n(y') = \hat{f}(y) - \hat{f}(y') + \rho \int \underbrace{[\hat{v}_{n-1}(y - u) - \hat{v}_{n-1}(y' - u)]dP(u)}_{\leq\ k}$$

for $y \leq y'$, hence

$$\hat{H}_n(y) - \hat{H}_n(y') \leq \hat{f}(y) - \hat{f}(y') + \rho k,\quad y \leq y'\ . \tag{42.8}$$

We set $y = s_n$, $y' = S_n$:

$$\underbrace{\hat{H}_n(s_n) - \hat{H}_n(S_n)}_{= k, \text{ i.e. order}} \leq \hat{f}(s_n) - \hat{f}(S_n) + \rho k$$

$$k(1 - \rho) \leq \hat{f}(s_n) - \hat{f}(S_n) .$$

Hence we order:

$$\boxed{\hat{f}(s_n) \geq \hat{f}(S_n) + (1 - \rho)k} \quad . \tag{42.9}$$

The above given inequality due to VEINOTT makes it possible to attribute the relationship between s_n and S_n, given by the result $\Delta H_n = k$ of the recursive function H_n, to the difference $\Delta \hat{f} = (1 - \rho k)$ resulting from n independent costs f.

$$\Delta \hat{H}_n = k \quad \Rightarrow \quad \Delta \hat{f} = (1 - \rho)k \qquad \text{for } s_n, S_n .$$

Hence an upper bound $\bar{s}$ is determined with given $\underline{S}$ from (42.9)

$$\boxed{\begin{array}{l} \bar{s} \text{ is the smallest number } < \underline{S} \text{ for which:} \\[2mm] f(\bar{s}) \leq f(\underline{S}) + (1 - \rho)k \quad . \end{array}} \tag{42.10}$$

The following figure shows this:

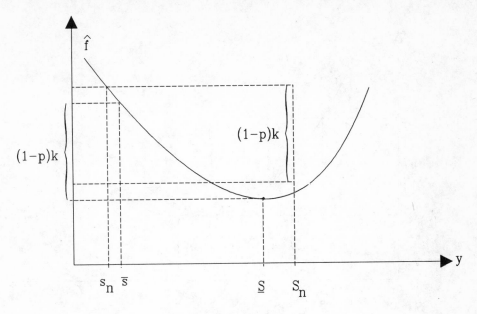

Figure 42.2: Boundary $\bar{s}$

Since $(1 - \rho)k > 0$ in the discounted case and $\hat{f}'(y) < 0$ for all $y < \underline{S}$, one also obtains the result

$$\boxed{\bar{s} < \underline{S} \ .}$$

(42.11)

3. Boundary for $\underline{s}$:

Consider $\hat{H}'_n(y) \leq \hat{f}'(y)$ for all $y < \underline{S}$, $n \in \mathbb{N}$ (compare (42.5)). From this it follows that

$$\hat{H}_n(y) - \hat{H}_n(\underline{S}) \geq \hat{f}(y) - \hat{f}(\underline{S}) \ , \quad y < \underline{S} \ ,$$

i.e., the level difference k is created if the interval $[y,\underline{S}]$ relative to $\hat{H}_n$ is smaller than that relative to $\hat{f}$. This is evident in Fig. 42.3:

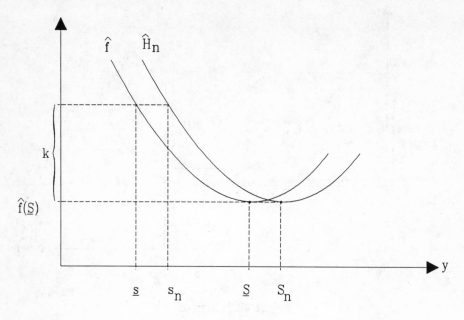

Figure 42.3: Lower bound $\underline{s}$

Hence, from the above diagram, $\underline{s}$ is the lower boundary for s_n

$\underline{s}$ is the smallest number $< \underline{S}$ for which:

$$\hat{f}(\underline{s}) \leq \hat{f}(\underline{S}) + k .$$

(42.12)

4. Bound $\tilde{S}$:

Initially an upper bound $\tilde{S}$ for D_n can be derived from: $f(\tilde{S}) = f(\underline{S}) + k$. A more precise bound, however, can still be found. Since the interval $[\underline{s}, \bar{s}]$ is defined by the difference ρk with respect to $\hat{f}$, the interval $[\underline{S}, \tilde{S}]$ must also be determined according to this difference with respect to $\hat{f}$. Hence we obtain $\tilde{S}$ from: $\hat{f}(\tilde{S}) = \hat{f}(\underline{S}) + \rho k$.

$$\boxed{\begin{array}{l} \overline{S} \text{ is the smallest number} > \underline{S} \text{ for which:} \\[2mm] \hat{f}(\overline{S}) \geq \hat{f}(\underline{S}) + \rho k \ . \end{array}}$$

(42.13)

Summary:

The parameters s_n, S_n are defined over the functions $\hat{H}_n$:

$$\hat{H}_n(S_n) = \underset{x}{\text{Min}} \ \hat{H}_n(x) \ \Rightarrow \ S_n \ ,$$

$$\hat{H}_n(s_n) = \hat{H}_n(S_n) + k \ \Rightarrow \ s_n \ .$$

Thus we order for all $y \leq S_n$ which satisfy the inequality

$$\hat{H}_n(y) - \hat{H}_n(S_n) \geq k \ .$$

The recursively defined functions $\hat{H}_n$, however, pose some difficulties. It is possible to estimate specific values of s_n, S_n, which were previously determined from the differences of H_n, from the differences in the single period cost $\hat{f}$. With this, the bounds $\underline{s}$, $\bar{s}$, $\underline{S}$, and $\overline{S}$ can be determined:

$\underline{s}$ = smallest whole number for which: $\hat{f}(\underline{s}) \leq \hat{f}(\underline{S}) + k$;

$\bar{s}$ = smallest whole number $< \underline{S}$ for which: $\hat{f}(\bar{s}) \leq \hat{f}(\underline{S}) + k(1 - \rho)$;

$\underline{S}$ = smallest whole number which minimizes $\hat{f}(\underline{S}) = \underset{y}{\text{Min}} \ \hat{f}(y)$;

$\overline{S}$ = smallest whole number $> \underline{S}$ for which: $\hat{f}(\overline{S}) \geq \hat{f}(\underline{S}) + \rho k$.

The graphical representation of these results are shown as follows:

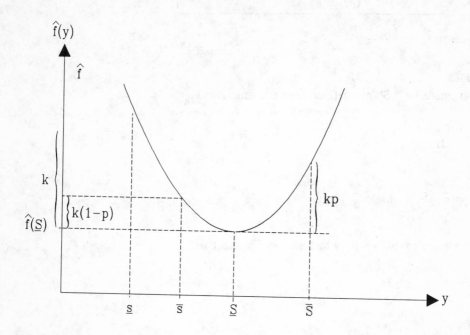

Figure 42.4: Boundaries of the (s_n, S_n) policy

As a by-product, the proof also gives the following results

$$\overline{s} < \underline{S},$$

$$\hat{f}(s_n) \geq \hat{f}(S_n) + (1 - \rho)k.$$

Remark: From the proof, it was assumed that a single minimum exists for f and $\hat{f}$ (at points S_1 and $\hat{S}_1$, respectively) and that the functions left of the minimum fall monotonically and, to the right of it, increase monotonically. Functions of this type are called <u>unimodal functions</u>.

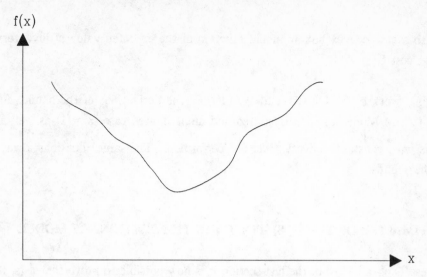

Figure 42.5: Graph of a unimodal function f(x)
with a minimum as extreme value

Thus the above results apply not only for models with convex single period costs (special case of unimodal) but also, in general, to unimodal costs. Strictly speaking inventory costs are not proportional to the quantity y. Only the interest cost is proportional to it. The handling cost, e.g. acquisition cost, increases at a decreasing rate with a more efficient inventory organization. The typical movement of the inventory cost now looks as follows:

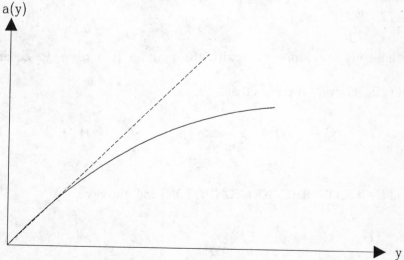

Figure 42.6: Returns to scale in inventory cost

This leads to a non–convex but unimodal function of the expected value of inventory and shortage cost $\hat{f}$.

In the original work by VEINOTT and WAGNER, the derivations of the bounds for s_n, S_n in an AHM Model for the discounted and undiscounted case ($\rho = 1$) as well as for the case with constant delivery time are performed. The same equations always apply for the bounds.

§43   OPTIMALITY OF THE (s, S)–POLICY IN THE STATIONARY MODEL

With the results obtained from the last section it is now possible to prove that a (s, S) policy is optimal for the stationary AHM Model with fixed costs. For the AHM Model without fixed ordering costs, one computes the optimal ordering rule of the (S) policy type by minimizing the single period cost. Since this does not depend on the planning horizon n, the (S) policy is also the optimal policy in the stationary model.

In an AHM Model with fixed ordering cost, the optimal ordering rule of the (s_n, S_n) policy type depends on the number of periods. Therefore it is more difficult to prove that a (s, S) policy is also optimal in the stationary case. We want to mention at this point only the essential steps (the proof is based on BANACH's fixed point theorem. See COLLATZ (1968)).

1.   The limit $v(y) := \lim_{n \to \infty} v_n(y)$ exists: the sequence $\{v_n\}$ converges equally at each finite interval. It follows that

$$\max_{\underline{s} \leq y \leq \overline{S}} \{|v_{n+1}(y) - v_n(y)|\} \leq \rho \max_{\underline{s} \leq y \leq \overline{S}} \{|v_n(y) - v_{n-1}(y)|\} \tag{43.1}$$

for all $n \in \mathbb{N}$ (CONTRACTION CONDITION) and, therefore, also

$$\max_{\underline{s} \leq y \leq \overline{S}} \{|v_{n+1}(y) - v_n(y)|\} \leq \rho^n \max_{\underline{s} \leq y \leq \overline{S}} \{|v_1(y)|\} . \tag{43.2}$$

It is enough to restrict the maximization within the area $\underline{s} < y < S$ since for $y < \underline{s}$, $v_n(y)$ is linear with slope $-a$ for all n and $y > S$ cannot occur in the stationary case.

2. Under the condition $v_0 = 0$, $\{v_n\}$ is a monotonically increasing sequence, i.e.,

$$v_{n+1}(y) \geq v_n(y) \quad \text{for all y, n} \; . \tag{43.3}$$

3. The function $v(y)$ satisfies the Principle of Optimality. We have

$$L(x,y,v) := a(x-y) + f(x) + \rho \int_0^\infty v(x-u) dP(u)$$

the right—hand side of BELLMAN's functional equation. Because of the monotonicity of the sequence $\{v_n\}$

$$v_n(y) = \underset{x \geq y}{\text{Min}} \; \{L(x,y,v_{n-1})\} \leq \underset{x \geq y}{\text{Min}} \; \{L(x,y,v)\} \; .$$

Hence, for $n \to \infty$

$$v(y) \leq \underset{x \geq y}{\text{Min}} \; L(x,y,v) \; . \tag{43.4}$$

On the other hand, due to the monotonicity of $\{v_n\}$ it follows that

$$v(y) \geq \underset{S \geq x \geq y}{\text{Min}} \; L(x,y,v_n)$$

$$\geq \underset{S \geq x \geq y}{\text{Min}} \; \{\underset{n \to \infty}{\lim} L(x,y,v_n)\} \; .$$

Using the LEBESQUE theorem of monotonic convergence, one may bring the limit value under the integral and obtain

$$v(y) \geq \underset{\bar{s} \geq x \geq y}{\text{Min}} \{L(x,y,v)\} = \underset{x \geq y}{\text{Min}} \{L(x,y,v)\} . \qquad (43.5)$$

Together with (43.4) we obtain

$$v(y) = \underset{x \geq y}{\text{Min}} \{L(x,y,v)\} . \qquad (43.6)$$

4. $v(y)$ is the unique solution of the BELLMAN functional equation (43.6).

5.   The sequences $\{s_n\}$, $\{S_n\}$ are restricted within $[\underline{s}, \bar{s}]$, $[\underline{S}, \bar{S}]$. These contain convergent partial sequences. Each limit s and S of these partial sequences describes an optimal ordering rule for the stationary model.

§44   A METHOD FOR COMPUTING s AND S

Discounted Case

We write the Principle of Optimality in the form:

$$\hat{v}(y) = \begin{cases} \hat{f}(y) + \rho \int\limits_0^\infty \hat{v}(y-u)dP(u), & \text{for} \quad y > s; \\ k + \hat{f}(S) + \rho \int\limits_0^\infty \hat{v}(S-u)dP(u), & \text{for} \quad y \leq s. \end{cases}$$

The functional equation defined for $\hat{v}$ can be derived iteratively as follows (let the initial inventory be S):

$$\hat{v}(S) = \underset{\text{in Period 1}}{\text{Expected Cost}} + \underset{\text{in Period 2}}{\text{Expected Cost}} + \underset{\text{in Period 3}}{\text{Expected Cost}} + \dots .$$

As long as the stock does not drop to nor below s, i.e., as long as the accumulated demand does not reach nor exceed the amount D, only the expected inventory and shortage costs $\hat{f}$ are incurred.

The fixed ordering cost k and the subsequent costs $\hat{v}(S)$ are incurred at $y \leq s$. Let

$p_x^{(n)}$:   Probability that the demand $u = x$ occurs exactly within n periods;

$P^{(n)}(x)$:   Distribution function of demand in n periods, i.e., the probability that the demand $u < x$ occurs at most within n periods.

$P(n)$ is the n–fold convolution of P

$$P^{(n)}(D) = \int_0^D P(D - u)\,dP^{(n-1)}(u).$$

Furthermore, let

$Q^{(n)}(D)$:   Probability that the accumulated demand reaches at least the amount D exactly in the $(n+1)$th period

$$Q^{(n)}(D) = \int_0^D [1 - P(D - u)]\,dP^{(n-1)}(u). \tag{44.1}$$

Then

$$\hat{v}(S) = \underbrace{\hat{f}(S) + \rho \int_0^D \hat{f}(S - u)\,dP(u) + \rho^2 \int_0^D \hat{f}(S - u)\,dP^{(2)}(u) + \dots}_{y \text{ stays above s in Periods } 1,2,3,\dots}$$

$$\underbrace{+ \rho\,[k + \hat{v}(S)]\, Q^{(1)}(D) + \rho^2[k + \hat{v}(S)]Q^{(2)}(D) + \dots}_{y \leq s \text{ in Periods } 1,2,3,\dots}$$

Using the following definition for the null convolution

$$p_u^{(o)} \, du := \begin{cases} 1 \, du, & \text{for } u = 0; \\ 0, & \text{for } u > 0; \end{cases} \tag{44.2}$$

we have

$$\hat{f}(S) = \rho^o \int_0^D \hat{f}(S) dP^{(o)} u \, ,$$

and obtain

$$\hat{v}(S) = \sum_{n=0}^\infty \rho^n \int_0^D \hat{f}(S-u) dP^{(n)}(u) + [k + \hat{v}(S)] \sum \rho^n q^{(u)}(D) \, . \tag{44.3}$$

Under the assumption that it will be ordered before shortage occurs, i.e.,

$$\hat{v}(0) = k + \hat{v}(S)$$

(44.3) can be solved for v(0),

$$\hat{v}(0) = \frac{k + \sum_{n=0}^\infty \rho^n \int_0^D \hat{f}(S-u) dP^{(n)}(u)}{1 - \sum_{n=1}^\infty \rho^n q^{(n)}(D)} \, .$$

We now reduce the terms $Q^{(n)}(D)$ in the denominator, $n = 1, 2, 3 \ldots$, into the distribution functions $P^{(n)}(D)$. From (44.1) we have

$$q^{(n)}(D) = \int_0^D dP^{(n-1)}(u) - \int_0^D P(D-u) dP^{(n-1)}(u)$$

$$= P^{(n-1)}(D) - P^{(n)}(D) \, .$$

Therefore the denominator becomes

$$1 - \sum_{n=1}^{\infty} \rho^n q^{(n)}(D) = 1 - \sum_{n=0}^{\infty} \rho^{n+1} [P^{(n)}(D) - P^{(n+1)}(D)]$$

$$= 1 - \rho P^{(0)}(D) - \sum_{n=1}^{\infty} P^{(n)}(D) \, (\rho^{n+1} - \rho^n)$$

$$= 1 - \rho P^{(0)}(D) + (1 - \rho) \sum_{n=1}^{\infty} \rho^n P^{(n)}(D)$$

$$= (1 - \rho) \sum_{n=0}^{\infty} \rho^n P^{(n)}(D) \quad ,$$

and for $v(0)$ we obtain

$$\hat{v}(0) = \frac{k + \displaystyle\sum_{n=0}^{\infty} \rho^n \int_0^D \hat{f}(S - u) dP^{(n)}(u)}{(1 - \rho) \displaystyle\sum_{n=0}^{\infty} \rho^n \int_0^D dP^{(n)}(u)} . \tag{44.4}$$

Minimization of the objective function

$$\underset{S,D}{\text{Min}} \; \hat{v}_{S,D}(0)$$

results in the optimal values S and D, $s = S - D$, of the stationary (s, S) policy.

Undiscounted Case

In the undiscounted case, $\hat{v} := \lim\limits_{n \to \infty} \hat{v}_n$ is known to be unbounded. The average cost per period, C or c, now enters as an optimization criterion instead of the present value of total cost if one ignores the proportional ordering cost. In §21, it was shown that the following relationship arises between these two criteria (21.3)

$$c = \lim_{\rho \to 1} (1 - \rho)\hat{v}_\rho \, ,$$

$$C = \lim_{\rho \to 1} (1 - \rho) v_\rho \, .$$

Applying this to the objective function $\hat{v}_\rho$ in (44.4) one obtains the cost criterion for the undiscounted case

$$c = \frac{k + \sum\limits_{n=0}^{\infty} \int_0^D \hat{f}(S - u) dP^{(n)}(u)}{\sum\limits_{n=0}^{\infty} \int_0^D dP^{(n)}(u)} \, . \tag{44.5}$$

Undiscounted Case with Discrete Demand

Previously, if we had used average costs per period (i.e., per unit time) in the stationary case, then the state probabilities were always considered. We again have a similar case. However, it is not obvious from the above objective function (44.5). We can reformulate it in the case of discrete demand so that only the state probabilites arise instead of the convolutions.

Firstly, (44.5) is formulated for the discrete case. Let

$p_u^{(n)}$: Probability that the demand is u over n periods;

$P^{(n)}(D)$: Probability that the demand is smaller than D over n periods.

With these expressions, (44.5) becomes

$$c = \frac{k + \sum\limits_{u=0}^{D-1} \hat{f}(S-u) \sum\limits_{n=0}^{\infty} p_u^{(n)}}{\sum\limits_{n=0}^{\infty} P^{(n)}(D)} \ . \tag{44.6}$$

The state probabilities π_y are now to be determined.

π_y :   Probability that the initial stock has the value y in the stationary case

After placing an order, the initial stock between s and S may lie in the range $s + 1 \le y \le S$.

We first consider the situation $y = S$. It can only exist if either the initial stock in the previous period was already S and no demand occurred

$$\pi_S p_0 \ ,$$

or demand has occurred in the previous period and because of it the stock fell to or below s; in this case, the probability is

$$\sum\limits_{y=s+1}^{S} \pi_y [1 - P(y-s)] \ .$$

Since both of these events are independent of each other, the state probability π_S is the sum of both probabilities above

$$\tau_S = \tau_S p_0 + \sum_{y=s+1}^{S} \tau_y [1 - P(y - s)]. \tag{44.7}$$

We now calculate the state probabilities τ_y, $y < S$.

We assume that $S - s \geq 2$. An initial stock $y < S$ may exist only in this case. We obtain the initial stock $y < S$ from the initial stock $y + u$ of the previous period and the demand u.

$$\tau_y = \sum_{i=y}^{S} \tau_i p_{i-y} \; ; \quad \text{for } s < y < S \; . \tag{44.8}$$

The solution of these equations (44.7) and (44.8) is obtained by the following consideration. The state $y < S$ is reached from S for the first time after n periods with probability

$$p_{S-y}^{(n)} \; , \quad \text{for } 1 \leq y < s \; .$$

Hence,

$$\tau_y = \tau_S \sum_{n=1}^{\infty} p_{S-y}^{(n)}, \quad \text{for } s < y < S \; . \tag{44.9}$$

With

$$p_u^{(0)} = \begin{cases} 1, \text{ for } u = 0 \; ; \\ 0, \text{ for } u > 0 \; ; \end{cases}$$

the summation can start at $n = 0$ and using the normalizing condition

$$\sum_{y=s+1}^{S} \pi_y = 1$$

we obtain the solution

$$\pi_y = \pi_{S-u} = \frac{\sum\limits_{n=0}^{\infty} p_u^{(n)}}{\sum\limits_{n=0}^{\infty} P^{(n)}(D)} \quad , \text{ for } u = 0,\dots,s. \tag{44.10}$$

We have

$$\sum_{n=0}^{\infty} p_0^{(n)} = \frac{1}{1 - p_0} \quad ,$$

so that π_S can simply be written as

$$\pi_S = \frac{1}{1 - p_0} \cdot \frac{1}{\sum\limits_{n=0}^{\infty} P^{(n)}(D)} \quad . \tag{44.11}$$

If one substitutes both of these results in the objective function (44.6), one obtains the formula

$$\boxed{c = k(1 - p_0)\pi_S + \sum_{y=s+1}^{S} \pi_y \hat{f}(y)} \tag{44.12}$$

Special Case: Geometric Demand Distribution

For a geometric demand distribution

$$p_u = qp^u , \quad 0 < p < 1 , \quad q = 1 - p , \quad u \in \mathbb{N}_0 .$$

The generating function (compare §19) is

$$G(x)_{geom} = \frac{q}{1 - px} ,$$

and the generating function of the n–fold convolution is

$$[G(x)]^n = (\frac{q}{1 - px})^n .$$

We now want to compute the state probabilities. By definition

$$\sum_{u=0}^{\infty} p_u^{(n)} x^u = [G(x)]^n .$$

If one takes the summation over all convolutions, $n = 1, 2, 3, \ldots$ with this equation, then one obtains

$$\sum_{n=1}^{\infty} \sum_{u=0}^{\infty} p_u^{(n)} x^u = \sum_{n=1}^{\infty} [G(x)]^n$$

$$= \frac{1}{1 - G(x)} - 1$$

$$= \frac{1}{1 - \frac{q}{1 - px}} - 1$$

$$= \frac{q}{p} \frac{1}{1 - x} ,$$

therefore

$$\sum_{u=0}^{\infty} \sum_{n=1}^{\infty} p_u^{(n)} x^u = \sum_{u=0}^{\infty} \frac{q}{p} x^u \ . \tag{44.13}$$

Comparing coefficients, we have

$$\sum_{n=1}^{\infty} p_u^{(n)} = \frac{q}{p} = \text{constant} \tag{44.14}$$

Substituting into (44.9) results in

$$\pi_y = \pi_S [1 - p_0] \frac{q}{p}$$

$$\pi_y = \pi_S q = \text{constant} \ , \qquad s < y < S. \tag{44.15}$$

From the normalizing condition

$$1 - \pi_S = \sum_{y=s+1}^{S-1} \pi_y$$

one obtains

$$\pi_S = \frac{1}{1 + (D - 1)q} \tag{44.16}$$

$$\pi_y = \frac{q}{1 + (D - 1)q} \ , \qquad s < y < S. \tag{44.17}$$

The objective function (44.12) then becomes

$$c = \frac{k(1 - q) + \hat{f}(S)}{1 + (D - 1)q} + \frac{q}{1 + (D - 1)q} \sum_{y=s+1}^{S-1} \hat{f}(y) \ .$$

We define the inventory and shortage costs $\hat{f}$ as

$$\hat{f}(y) = hy + g \sum_{u=y+1}^{\infty} (u-y)p_u \, . \tag{44.18}$$

The function $f(y)$ is convex so that an optimal (s, S) policy remains guaranteed. However, it estimates the inventory cost in a most unfavorable way.

With the geometric distribution we have

$$f(y) = hy + \frac{gp^{y+1}}{q} \, .$$

We substitute this value in the objective function and obtain

$$c = \frac{kp + hS + g\,\dfrac{p^{S+1}}{q} + qh \displaystyle\sum_{y=s+1}^{S-1} y + qg \displaystyle\sum_{y=s+1}^{S-1} \dfrac{p^{y+1}}{q}}{1 + (D-1)q}$$

$$c = \frac{kp + h[S + q(D-1)s + qD(\frac{D-1}{2})] + g[\frac{p^{S+1}}{q} + \frac{p^{S+2} - p^{S+1}}{1-p}]}{1 + (D-1)q}$$

$$c = h(s + \frac{D}{2}) + \frac{kp + h\,\dfrac{D}{2} + g\,\dfrac{p^{s+2}}{q}}{1 + (D-1)q} \, . \tag{44.19}$$

Determination of s:

s is the smallest number for which the first difference of c in s is greater than or equal to zero:

$$\Delta_s c = h + \frac{g \, \frac{p^s}{q} \, (p-1)}{1 + (D-1)q} \, .$$

This is the same as

$$p^s \geq \frac{h}{g} \, [1 + (D-1)q] > p^{s+1} \, . \tag{44.20}$$

We then obtain

$$s = \frac{\log \frac{h}{g} \, [1 + (D-1)q]}{\log p} \, . \tag{44.21}$$

Determination of D:

D is the smallest whole number for which the first difference of c in D is greater than or equal to zero. After some computational steps one obtains

$$h \left[\frac{q^2 D}{2} \, (D-1) + q(D-1) + 1 \right] \geq qpk + gp^{s+1} \, .$$

If one substitutes p^{s+1} by the approximation (44.20), then one obtains

$$D(D-1) = \frac{2pk}{qh} \, .$$

Noting that p/q is precisely the expected value μ for demand, we then have

$$D(D-1) = \frac{2k\mu}{h} \, . \tag{44.22}$$

This result is similar to the Wilson Formula $D = \sqrt{\frac{2k\mu}{h}}$ for the case with deterministic demand rate μ.

The above results are based on the fact that the state probabilities π_y for $y < S$ are identical (see (44.17)). This is correct only for the geometric and the binomial distributions. If μ is small, then the state probabilities π_y are also almost identical for other distributions distributions and the formulas (44.21) and (44.22) serve as good estimates.

§45   AHM–MODEL WITH DELIVERY TIME

We previously assumed that the stock at the beginning of a period can be replenished immediately. This is an ideal which can only be justified for long periods. As a rule, one must reckon with delivery times even if the goods come from the same firm. We describe delivery as encompassing a number of individual activities from dispatching, to loading, transporting, unloading, receiving inspection up to storage, all of which require a certain amount of time.

We therefore generalize the AHM Model for the case without reliable delivery times. It will be shown that the expected model does not go beyond the framework of the AHM–Type. It is, however, only possible (with only a single exception) at the cost of increasing the state space by at least one dimension. Let

τ :   Delivery time

Basic for all models with delivery time is the consideration that costs are related to the point in time when the ordered quantity is eventually delivered. One does not have any influence on the status of inventory before this time because a present action only has an effect after τ periods.

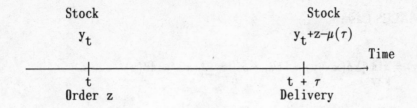

Figure 45.1

The quantity in the above figure $y_{t+\tau} = y_t + z - u(\tau)$ is correct if one adds the on–order quantities to the physical stock. This is reasonable, since these quantities are available at time $t + \tau$, while they are still on–order at time τ. Hence, we describe the models with delivery time

y :   Stock on–hand plus on–order
z :   Order quantity

Case 1: Delivery Time $\tau = 1$
The order quantity is delivered at the start of the next period.

BACKORDER Case:

$$\hat{v}_n(y) = \underset{z \geq 0}{\text{Min}}\ \{k\delta(z) + \hat{f}(y) + \rho \int_0^\infty \hat{v}_{n-1}(y + z - u)dP(u)\}.$$

Since $z = x - y$, it then becomes

$$\hat{v}_n(y) = \underset{x \geq y}{\text{Min}}\ \{k\delta(x - y) + \hat{f}(y) + \rho \int_0^\infty \hat{v}_{n-1}(x - u)dP(u)\}. \qquad (45.1)$$

LOST–SALES Case:

$$\hat{v}_n(y) = \underset{x \geq y}{\text{Min}} \{ k\delta(x-y) + \hat{f}(y) + \rho \int_0^y \hat{v}_{n-1}(x-u)dP(u)$$

$$+ \rho \hat{v}_{n-1}(x-y)[1-P(y)]\} . \qquad (45.2)$$

It is to be noted here that shortages may occur even before an order arrives. Therefore, shortage is given by $y - u < 0$ and not by $x - u < 0$. Since $x \geq y$, shortages occur with a greater probability as compared to the model with delivery time $\tau = 0$

$$1 - P(y) \geq 1 - P(x) .$$

This implies that the stock fluctuation becomes larger for models with delivery time as compared to models without delivery time. This also applies if the delivery time is fixed, i.e., reliable. The delivery time makes inventory more expensive (i.e., if $g > h$).

Case 2: $\tau = 2$.

If the delivery time lasts for two periods, one has to take note of the quantity of the previous period. Let

y	: Stock on–hand plus the remaining order quantity two periods before
z_1	: Order quantity from the previous period
z	: Actual order quantity, $z = x - y$.

The decision about the present order quantity depends on the value of both states y and z_1. The principle of optimality in the BACKORDER case is

$$\hat{v}_n(y,z_1) = \underset{z}{\text{Min}} \{ k\delta(z) + \hat{f}(y) + \rho \int_0^\infty \hat{v}_{n-1}(y + z_1 - u, z)dP(u)\} , \qquad (45.3)$$

in the LOST–SALES case

$$\hat{v}_n(y,z_1) = \underset{z}{\text{Min}} \{k\delta(z) + \hat{f}(y) + \rho \int_0^y \hat{v}_{n-1}(y + z_1 - u, z)\,dP(u)$$

$$+ \rho\,\hat{v}_{n-1}(z_1,z)\,[1 - P(y)]\} \; . \qquad (45.4)$$

Case 3: $\tau = m$.

In general, let the delivery time τ be m periods long, m ϵ N.

Let

y   : Stock on–hand plus the remaining order quantity of the previous m^{th} period

z_i : Order quantity of the previous i^{th} period

A single variable representing the total remaining orders of the previous periods is not enough to describe the state of the system. To be able to explain the development of the stock, $y_t \to y_{t+1}$, each single order, z_i, must first be noted and will be used only upon the delivery of the latest order z_m. We therefore have a vector of m states

$$(y,z_1,z_2,\ldots,z_{m-1}) \; .$$

These are written as a recursion according to the formula

$$y \to y + z_{m-1} - u \qquad \text{(Inventory Balance Equation)}$$

$$(z_1,z_2,\ldots,z_{m-1}) \to (z,z_1,\ldots,z_{m-2}) \; .$$

The Principle of Optimality is

in the BACKORDER– Case:

$$\hat{v}_n(y, z_1, \ldots, z_{m-1}) = \underset{z}{\text{Min}} \{ k\delta(z) +$$

$$+ \hat{f}(y) + \rho \int_0^\infty \hat{v}_{n-1}(y + z_{m-1} - u, z, z_1, \ldots, z_{m-2}) dP(u) \}, \qquad (45.5)$$

in the LOST SALES Case:

$$\hat{v}_n(y, z_1, \ldots, z_{m-1}) = \underset{z}{\text{Min}} \{ k\delta(z) +$$

$$+ \hat{f}(y) + \rho \int_0^y \hat{v}_{n-1}(y + z_{m-1} - u, z, z_1, \ldots, z_{m-2}) dP(u)$$

$$+ \rho \, \hat{v}_{n-1}(z_{m-1}, z, z_1, \ldots, z_{m-2}) [1 - P(y)] \}. \qquad (45.6)$$

Case 4: Delivery time τ is not discrete.

The delivery time τ need not be an integer. Nothing is changed in the BACKORDER–Case for $0 < \tau < 1$ and $m - 1 < \tau < m$. The formulas (45.1), (45.3) and (45.5) are still applicable. We just round off the delivery time. There is, however, a difference in the LOST–SALES Case.

Since the last pending delivery has already arrived before the end of the present period, it can still be added to the saleable stock. Then, instead of (45.6), it must now be

$$\hat{v}_n(y, z_1, \ldots, z_{m-1}) = \underset{z}{\text{Min}} \{ k\delta(z) + \hat{f}(y)$$

$$+ \rho \int_0^{y+z_{m-1}} \hat{v}_{n-1}(y + z_{m-1} - u, z_1, \ldots, z_{m-2}) dP(u)$$

$$+ \rho \, \hat{v}_{n-1}(0, z, z_1, \ldots, z_{m-2}) [1 - P(y + z_{m-1})] \}. \qquad (45.7)$$

The number of possible states for $\tau = 2$ is already large. Hence, an approximation is also needed here.

Approximation

We revise the period length so that it is identical to the delivery time. We then have a model with a delivery time of one (long) period. This rescaling of time changes the distribution of demand. For a delivery time of $\tau = m$ periods, the demand $u_1 + u_2 + \ldots + u_m$ occurs as total demand over m periods. Since demand in the individual periods is assumed to be stochastic as well as independent, the distribution function of total demand is the m–fold convolution of the distribution function $P(u)$.

$P^{(m)}(u)$:   Distribution function of demand within the period τ
               (for delivery time m).

The holding and shortage costs are then defined as

$$\hat{f}^{(m)}(y) = (\hat{h} + \hat{g}) \int_0^y P^{(m)}(u)\,du + \hat{g}(m\mu - x)$$

and the value function satisfies the recursion in the BACKORDER–Case:

$$\hat{v}_n(y) = \underset{x \geq y}{\text{Min}} \left\{ k\delta(x - y) + \hat{f}^{(m)}(y) + \rho \int_0^\infty \hat{v}_{n-1}(x - u)\,dP^{(m)}(u) \right\}, \tag{45.8}$$

and the LOST–SALES Case:

$$\hat{v}_n(y) = \underset{x \geq y}{\text{Min}} \left\{ k\delta(x - y) + \hat{f}^{(m)}(y) + \rho \int_0^y \hat{v}_{n-1}(x - u)\,dP^{(m)}(u) \right.$$

$$\left. + \rho\, \hat{v}_{n-1}(x - y)\,[1 - P^{(m)}(y)] \right\}. \tag{45.9}$$

For this approximation model, the following is to be noted: By lengthening the period, S, as well as D, will become larger. Thus the quantity–dependent costs increase. The fixed ordering cost is not affected but remains constant. Therefore, the lengthening of the period has the effect of relatively reducing the fixed ordering costs and should be checked in individual cases whether one is better off without the fixed ordering cost in the model. In case m is large and k is small in the original model, a model without fixed ordering cost is suggested as an approximation. In this case, a S(m) policy is optimal with

$$S^{(m)} = P^{(m)^{-1}} \left(\frac{\hat{g}}{\hat{h} + \hat{g}} \right) . \tag{45.10}$$

As much as possible, one should avoid lengthening the period because it increases the variance of demand, which again drives the inventory costs higher (compare §38).

Stochastic Delivery Time τ

We consider here the simplest case: only one order remains at the most. Let

q_τ :	Probability that the delivery time is equal to τ
$Q(\tau)$:	Distribution function of delivery time
$\hat{v}_n(y,\tau,z_1)$:	Value function with a planning horizon of n periods, stock y and remaining order quantities z1 since τ periods.

τ is now an additional state value.

For the functional equation of the value function, one needs the transition probabilities

$\varphi_{\tau+1}$:   Transition probability from the state "still remaining after τ periods" to the state "delivered in period $\tau+1$".

Then

$$\varphi_{\tau+1} = \frac{q_{\tau+1}}{Q(\tau)} , \qquad\qquad \tau > 0; \qquad\qquad (45.11)$$

$$\varphi_1 = q_1 , \qquad\qquad \tau = 0 . \qquad\qquad (45.12)$$

As long as an order is still in transit no new order may be placed. The next order is allowed only when $z_1 = 0$. Hence, the Principle of Optimality is now contained in both equations, for the BACKORDER Case,

$$\hat{v}_n(y,z_1,\tau) = \hat{f}(y) + \rho\varphi_{\tau+1} \int_0^\infty \hat{v}_{n-1}(y + z_1 - u,0,0)dP(u)$$

$$+ \rho[1 - \varphi_{\tau+1}] \int_0^\infty \hat{v}_{n-1}(y - u,z_1,\tau+1)dP(u) \qquad\qquad (45.13)$$

$$\hat{v}_n(y,0,0) = \underset{x \geq y}{\text{Min}} \{k\delta(x - y) + \hat{f}(y) + \rho\varphi_1 \int_0^\infty \hat{v}_{n-1}(x - u,0,0)dP(u)$$

$$+ \rho[1 - \varphi_1] \int_0^\infty \hat{v}_{n-1}(y - u,x - y,1)dP(u)\} \qquad\qquad (45.14)$$

and, for the LOST SALES Case,

$$\hat{v}_n(y,z_1,\tau) = \hat{f}(y) + \rho\varphi_{\tau+1} \int_0^y \hat{v}_{n-1}(y + z_1 - u,0,0)dP(u)$$

$$+ \rho\varphi_{\tau+1}[1 - P(y)]\hat{v}_{n-1}(z_1,0,0)$$

$$+ \rho[1 - \varphi_{\tau+1}] \int_0^y \hat{v}_{n-1}(y - u,z_1,\tau+1)dP(u)$$

$$+ \rho[1 - \varphi_{\tau+1}][1 - P(y)]\hat{v}_{n-1}(0,z_1,\tau+1) , \qquad\qquad (45.15)$$

$$\hat{v}_n(y,0,0) = \underset{x \geq y}{\text{Min}} \{k\delta(x-y) + \hat{f}(y)$$

$$+ \rho\varphi_1 \int_0^y \hat{v}_{n-1}(x-u,0,0)\,dP(u)$$

$$+ \rho\varphi_1[1-P(y)]\,\hat{v}_{n-1}(x-y,0,0)$$

$$+ \rho[1-\varphi_1] \int_0^y \hat{v}_{n-1}(y-u,x-y,1)\,dP(u)$$

$$+ \rho[1-\varphi_1][1-P(y)]\,\hat{v}_{n-1}(0,x-y,1)\} \qquad (45.16)$$

§46   AUTOCORRELATED DEMAND

The Planning Research Corporation of the US Navy has already accepted the AHM Model with constant expected value of demand. A slide rule whereby the (s, S) policy can be determined was developed by M.J. BECKMANN . The parameters are μ, g/h and k. (See §38: Standardization). Models whose solutions required a stochastic dynamic program were at that time computationally cumbersome. However, the assumption of a stationary demand distribution is very unrealistic for some applications. With the increasing performance of the computer, it will be easier to use more realistic but also more computationally intensive inventory models.

In this section we handle the case where the demand u is influenced by its past history. We define a conditional density function as

$$p(u)du = p(u|u_1,u_2,\ldots,u_k)du$$

u_i :   Demand in the previous ith period

We specially consider the MARKOV Case

$$p(u)du = p(u|u_1)du \ . \qquad (46.1)$$

The last observation u_1 is taken to be the state variable in the optimality equations. These are

$$v_n(y,u_1) = \min_{x \geq y} \{k\delta(x-y) + a(x-y) + f(x,u_1) +$$

$$+ \rho \int_0^\infty v_{n-1}(x-u,u)dP(u|u_1)\} \ , \ n = 1,2,\ldots , \tag{46.2}$$

whereby

$$f(x,u_1) = h \int_{-\infty}^{x} (x-u)dP(u|u_1) + g \int_{x}^{\infty} (u-x)dP(u|u_1) \ . \tag{46.3}$$

What does one obtain by including the last observation? To answer this question, let us look closer at the MARKOV Process. Let

1) $p(u|u_1)du = \Psi(u - \mu(u_1))$,

   i.e., the expected value μ is dependent on the last observation

2) μ linear: $\mu = \mu_0 + \alpha(u_1 - \mu_0)$

   μ_0 long–term average

One of the more frequently used demand processes is the resulting special case of the autocorrelated process of the first order, the so–called AR(1)–Process. The process equation is

$$u_t - \mu_0 = \alpha(u_{t-1} - \mu_0) + \epsilon_t , \tag{46.4}$$

where $|\alpha| < 1$, ϵ_t for all t independent and identically $(0, \sigma_\epsilon)$–normally distributed with distribution function $\Psi(\epsilon)$. Then

$$u_t = \alpha^k(u_{t-k} - \mu_0) + \sum_{i=0}^{k-1} \alpha^i \epsilon_{t-i} + \mu_0 \; .$$

In the steady state, i.e., for $t \to \infty$

$$u_t = \sum_{i=0}^{\infty} \alpha^i \epsilon_{t-i} + \mu_0 \; , \qquad (46.5)$$

$$E\{u_t\} = \frac{1}{1-\alpha} E\{\epsilon_t\} + \mu_0 = \mu_0 \; ,$$

$$\sigma_u^2 = E\{(u_t - \mu_0)^2\}$$

$$= E\{[\sum_{i=0}^{\infty} \alpha^i \epsilon_{t-i} + \mu_0 - \mu_0]^2\}$$

$$= \frac{1}{1-\alpha^2} \sigma_\epsilon^2$$

$$\Rightarrow \quad \sigma_u^2 > \sigma_\epsilon^2 \quad \text{for} \quad |\alpha| < 1 \; . \qquad (46.6)$$

Note that for $|\alpha| < 1$, the unconditional process $\{u_t\}$ is also stationary with $E\{u_t\} = \mu_0$ and $E\{[u_t - \mu_0]^2\} = \sigma_u^2$. But by including the last observation u_1 in memory, i.e., according to the formulation of the AR(1) Process, the variance of demand becomes smaller in the present period as was shown in (46.6).

§47   INVENTORY WITH FORECASTING

The introduction of the AHM inventory model can sometimes fail in models which imply a stationary demand process. However, information about the future course of demand usually exists from which short–term forecasts can be made. These forecasts must be considered in the model. One finds such situations, for example, if supply agreements are settled with a major client.

In practice, one proceeds in such a way that a demand forecast is made in the first step, a safety stock level is determined in the second step, the forecast as deterministic demand is determined in the third step and, using a deterministic model, the optimal ordering rule is computed. This stepwise procedure, however, leads to solutions which are, as a rule, sub–optimal.

Optimal solutions are obtained if the forecasts are integrated in the dynamic programming method. This requires a reformulation of the optimality principle.

We differentiate two types of forecasts: the exogeneous and the endogeneous forecasts. In the endogeneous forecast, the forecast values are solely derived from the previously observed demand. The autoregressive scheme is a special case. We summarize the preceding observations u_1, u_2, ... in a sufficient statistic and compute using the conditional density

$$p(u|w_1)$$

instead of the conditional density

$$p(u|u_1,u_2,\ldots) \ .$$

The information extracted from the past values u1, u2, ... are considered as a forecast for the demand u occurring in the present period. If the exact value of u is known at the end of the period, a new forecast can be computed with the help of the forecast formula

$$w = g(u, w_1) \tag{47.1}$$

It is important that the new forecast w can be derived recursively from the old forecast w_1 and the latest observation of demand u. It is then possible to integrate the forecasting process in the optimality principle. It is given as

$$v_n(y, w_1) = \underset{x}{\text{Min}} \ \{k\delta(x-y) + a(x-y) + f(x, w_1) +$$

$$+ \rho \int_0^\infty v_{n-1}(x-u, w) \, dP(u|w_1)\} \ . \tag{47.2}$$

Example: First Order Exponential Smoothing

The sufficient statistic w is a weighted average of all observations u_1, u_2, ... whereby the observations preceding k periods are weighted by the factor α^k, $0 < \alpha < 1$:

$$p(\underbrace{u|u_1, u_2, \ldots}_{})$$

$$w_1 = (1-\alpha) \sum_{k=0}^\infty u_{k+1} \alpha^k \tag{47.3}$$

The forecast formula is (compare §6)

$$w = \alpha w_1 + (1-\alpha)u \ . \tag{47.4}$$

If one formulates it depending on t, i.e.,

$$w_{t+1} = \alpha w_t + (1-\alpha)u_{t+1} \ ,$$

then one observes that in the stationary state

$$w = u$$

and, hence, w can be reasonably interpreted as a forecast of a non–periodic stationary process.

The forecast is unbiased for the stationary demand process $\{u_t\}$:

$$E\{w\} = (1 - \alpha) \sum_{k=0}^{\infty} \alpha^k E\{u_{t-k-1}\} = \mu = E\{u\}$$

and has a variance

$$Var\{w\} = (1 - \alpha)^2 \sum_{k=0}^{\infty} \alpha^{2k} \sigma_u^2$$

$$= \frac{(1 - \alpha)^2}{1 - \alpha^2} \sigma_u^2 < \sigma_u^2$$

which is smaller than σ_u^2.

Conditional Expected Value as Forecast

Indeed, the above forecast is unbiased but does not have a minimum variance. Forecasts with minimal variance are obtained if one chooses the conditional expected value as the forecast formula which is the case in an autoregressive scheme.

Exogeneous Forecasts

In the exogeneous forecast, the source of information lies outside of the model. This situation exists, for example, if the forecast and stock control functions are done in different departments of a firm. The forecasting department makes its forecast data based on business and economic data. For the inventory manager, the forecast takes on the character of a random variable:

w_1 The latest forecast. It is based on the demand of the previous period.

u   has the density $p(u|w1)$ du.

w   is the forecast still to be made for the future period. w is, in the eyes of the inventory manager, a random variable, which is hidden from him by the forecast mechanism.

$\varphi(w)dw$ is the density of w.

The dynamic program is then

$$v_n(y,w_1) = \underset{x \geq y}{\text{Min}} \{k\delta(x-y) + a(x-y) + f(x, w_1)$$
$$+ \rho \iint v_{n-1}(x-u,w)p(u|w_1)\Psi(w)du\ dw\}. \tag{47.5}$$

The advantage of this model as compared to the model without a forecast is that w_1 now allows the distribution of demand to concentrate more on the short–term expected value (in case the forecast is good!). This advantage, however, can again be partially lost since w introduces new uncertainties into the model. This is expressed in the double integral in (47.5). The double integration smooths the cost difference between the favorable and the unfavorable states. The cost curve v becomes flatter.

Reduction of the State Space

Under the two conditions

C1) $p(u|w_1) = \varphi(u-w_1) = \varphi(\epsilon)$ with constant variance

C2)   w independent of u and w_1

the dynamic program can be formulated with only one state variable.
We rewrite the Principle of Optimality in which we use the variable

$$\epsilon := u - w_1 \quad (= \text{Forecast error})$$

i.e.,

$$v_n(y,w_1) = \underset{x \geq y}{Min} \{k\delta[x - w_1 - (y - w_1)] + a[x - w_1 - (y - w_1)] +$$

$$+ h \int_{-\infty}^{x-w_1}(x - w_1 - \epsilon)\varphi(\epsilon)d\epsilon + g \int_{x-w_1}^{\infty}[\epsilon - (x - w_1)]\varphi(\epsilon)d\epsilon$$

$$+ \rho \iint v_{n-1}[x - w_1 - \epsilon, w]\varphi(\epsilon)\Psi(w)d\epsilon \, dw\} \, . \tag{47.6}$$

We consider the new state values

$$r := y - w_1 \, ; \tag{47.7}$$

$$\xi := x - w_1 \, ; \tag{47.8}$$

where r and ξ are net inventories, i.e., inventories which exclude the estimate w_1 of demand u.

r :   Net beginning inventory before the order

ξ :   Net beginning inventory after the order, i.e., $\xi = r + z$

With these new state values, (47.6) becomes

$$v_n(y,w_1) = \underset{\xi \geq r}{Min} \{k\delta(\xi - r) + a(\xi - r)$$

$$\underbrace{y - w_1}_{r}$$

$$+ h \int_{-\infty}^{\xi}(\xi - \epsilon)\varphi(\epsilon)d\epsilon + g \int_{\xi}^{\infty}(\epsilon - \xi)\varphi(\epsilon)d(\epsilon)$$

$$+ \rho \iint v_{n-1}\underbrace{[\xi - \epsilon, w]}_{\xi - \epsilon - w}\varphi(\epsilon)\Psi(w)d\epsilon \, dw\} \, . \tag{47.9}$$

The right–hand side no longer depends on w_1. Therefore, one can give up the second state variable w_1 and the Principle of Optimality can be formulated into a single state variable as "net inventory".

$$
\begin{aligned}
v_n(r) = \min_{\xi \geq r} \{ & k\delta(\xi - r) + a(\xi - r) \\
& + h \int_{-\infty}^{\xi} (\xi - \epsilon)\varphi(\epsilon)d\epsilon + g \int_{\xi}^{\infty} (\epsilon - \xi)\varphi(\epsilon)d(\epsilon) \\
& + \rho \iint v_{n-1}(\xi - \epsilon - w)\varphi(\epsilon)\bar{\Psi}(w)d\epsilon \ dw \} \ .
\end{aligned}
$$

(47.10)

How realistic are these two conditions C1 and C2?

First, the normal distribution satisfies C1, since

$$
n(u|w_1) = \frac{1}{\sqrt{2\pi}\ \sigma}\ e^{-\frac{1}{2\sigma^2}(u - w_1)^2} = n(u - w_1) \ .
$$

A constant variance σ_u^2 is also required in C1. As a rule, this is accepted for goods with a very low market growth. For a higher market growth, σ_u^2 would also increase.

Second, the forecasts w and w_1 should not be autocorrelated.

Third, the demand u should not influence the forecast. Therefore, goods whose output are representative of a key industry are excluded.

For example, if w is the change in the gross national product, then the above method is applicable for goods which follow the acceleration principle, e.g., investment goods and spare parts.

Exogenous Forecast with Self–adaptation

For certain goods which do not fulfill the above conditions, the forecast w will depend on the old forecast w_1 and on the immediate demand u. w then has a conditional density

$$\Psi(w|w_1,u)dw \ .$$

The Principle of Optimality can now be written in the form

$$v_n(y,w_1) = \underset{x \geq y}{\text{Min}} \ \{k\delta(x-y) + a(x-y) +$$

$$+ h \int_{-\infty}^{x} (x-u)p(u|w_1) + g \int_{x}^{\infty} (u-x)p(u|w_1) + \qquad (47.11)$$

$$+ \rho \iint v_{n-1}(x-u,w)p(u,w_1)\Psi(w|w_1,u)du \ dw\} \ .$$

In this manner, an adaptive mechanism is integrated in the dynamic program.

CHAPTER 6:
NUMERICAL METHODS

In the preceding chapters, a number of basic inventory models – especially the stochastic models – were introduced. Most of these models, however, must be modified to fit the requirements in actual practice. Because of this, the models may change in such a way that the suggested solution is no longer suitable; for example, when considering multiple discount breaks and transportation costs. This chapter, therefore, introduces a number of numerical methods which help one to calculate very general models. The method due to FEDERGRUEN and ZIPKIN (1984) discussed in the last section is an exception.

§48   VALUE ITERATION

All inventory models, which can be formulated using Bellman's Principle of Optimality, can be solved by the value iteration method. It involves the recursive evaluation of the functional equation of dynamic programming.

General Method of Value Iteration

Step 1:   Start with $v_0 \equiv 0$ or another vector appropriate to the problem, $n = 1$ and a termination criterion.

Step 2:   Compute v_n using the principle of optimality.

Step 3:   Termination criterion satisfied?

                No : Set $n := n + 1$ and go to 2.

                Yes: Go to 4.

Step 4:   Stop.

A termination criterion can be set,

a)   in the case of a finite planning horizon, upon reaching the end of the horizon

b)   in the case of an infinite planning horizon:

   – a maximum number of iterations is reached

   – when an absolute increase falls below a given minimum

$$| v_{n+1} - v_n | < \epsilon_{abs}$$

   – when a relative increase falls below a given minimum

$$| v_{n+1} - v_n | / | v_{n+1}| < \epsilon_{rel}$$

   – when the difference between successive increases falls below a given minimum (specially in an undiscounted case)

$$| \, | v_{n+1} - v_n | - | v_n - v_{n-1} | \, | < r \, .$$

Infinite planning horizon

The following considerations assume an infinite horizon.

Inventory models with identical demand distributions in each period can be formulated as a homogeneous Markovian decision process. For numerical purposes, we assume the discrete case.

Let

i	:	State, i = 1,2,...N
d	:	Decision, d_i is the decision in state i
δ	:	Ordering rule $\delta = (d_1, ..., d_n)$; (δ is no longer the Kronecker symbol!)
p_{ij}^d	:	Transition probability from i to j with decision i
a_i^d	:	Expected value of the single period cost, depending on state i with decision d_i

We set costs as negative values and obtain a maximization problem

$$v_n(i) = \max_d \left\{ a_i^d + \rho \sum_{j=1}^N p_{ij}^d v_{n-1}(j) \right\} , \qquad (48.1)$$

n = 1, 2, 3, ..., v_o given, $\rho \leq 1$.

or in vector notation

$$v_n = \max_\delta \left\{ a_\delta + \rho P_\delta v_{n-1} \right\} \qquad (48.2)$$

with

$$a_\delta = \begin{bmatrix} a_1^{d_1} \\ a_2^{d_2} \\ \vdots \\ a_N^{d_N} \end{bmatrix} ; \quad P_\delta = \begin{bmatrix} p_{11}^{d_1} & p_{12}^{d_1} & P_{1N}^{d_1} \\ \vdots & & \\ P_{N1}^{d_N} & P_{N2}^{d_N} & P_{NN}^{d_N} \end{bmatrix} .$$

In a (s,S)–ordering rule, P_δ has the following form

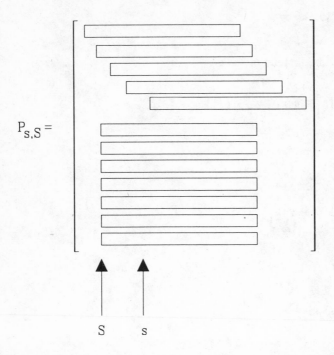

The bars ⬚ describe the distribution $(p(1), p(2),...p(u_{max}))$ of demand u. All other elements of the matrix are zero.

For simplification we introduce the following notations

a)   in case the decision in each state is fixed and not maximized then the function w replaces the value function v,

$$w_n(i) = a_i^d + \rho \sum_j p_{ij}^d w_{n-1}(j) \ ,$$

and we shorten this equation as follows

$$w_n(i) = l(d,i,w_{n-1})$$

or in vector notation

$$w_n = L(\delta, w_{n-1}) \; ;$$

b)   by maximization, instead of

$$v_n(i) = \max_d \{a_i^d + \rho \sum_j p_{ij}^d v_{n-1}(j)\}$$

we now write

$$v_n(i) = \max_d l(d, i, v_{n-1}) \; ;.$$

or as a vector

$$v_n = \max_\delta L(\delta, v_{n-1}) \; ;$$

and using

$$U := \max_\delta L$$

the short form

$$v_n = U v_{n-1}$$

is obtained.

With this notation, the general method for value iteration in the discounted case, i.e., for $j < 1$ with an infinite planning horizon, is transformed into the following algorithm:

Value iteration in the discounted case, infinite planning horizon

Step 1:   Start with $v_0 (\equiv 0), \epsilon_{abs} > 0$.

Step 2:   Compute Uv (maximizing step).

Step 3: $\| Uv - v \| > \epsilon_{abs}$?

Yes: Set $v := Uv$ and go to 2.

No: Go to 4.

Step 4:   Stop.

The supremum norm $\| v \| = \max_i \{v(i)\}$ is used.

The above value iteration is sufficient for convergence if the largest absolute eigenvalue of the iteration matrix ρP is less than 1.

Lemma 48.1: The matrix ρP has a highest possible absolute eigenvalue $|\lambda|_{max} = \rho$.

Proof:
The matrix P has an eigenvector, $(1,...,1) =: e^T$, which corresponds to the eigenvalue 1. Since an upper bound of λ is given by

$$|\lambda| \leq \max_i \sum_j |p_{ij}| = 1 \text{ (Eigenvalue estimate)}$$

$\lambda = 1$ is the greatest eigenvalue of P. Hence ρ is the greatest eigenvalue of ρP. *q.e.d.*

Thus the convergence of value iteration in a discounted optimization problem is, therefore, assured.

Aside from this,

Lemma 48.2:   For $\rho < 1$, L is contracting, i.e., for all $u,v \in \mathbb{R}^N$ and for all δ,

$$| L(\delta,u) - L(\delta,v) | \leq \rho | u - v | , \quad 0 < \rho < 1 .$$

Proof: $| L(\delta,u) - L(\delta,v) | = | a_\delta + \rho P_\delta u - a_\delta - \rho P_\delta v |$

$$= | \rho P_\delta(u-v) \| \le \rho \| P_\delta \| \| u-v \|$$

$$= \rho | u - v | . \quad \text{q.e.d.}$$

Next, it will be shown that the contraction property also holds for U.

Lemma 48.3:   For $\rho < 1$, U is contracting, i.e., for all $u,v \in \mathbb{R}^N$

$$| Uu - Uv | \le \rho | u - v |, \quad 0 < \rho < 1.$$

Proof:

Let i be any component of vector v (i.e., we take any state). Then for $u,v \in \mathbb{R}^N$: $(Uu)_i$ $= (Uv)_i + k$. Let $k > 0$ and let $\tilde{d}$ be the maximizing decision with respect to u in i, i.e.,

$$(Uu)_i = l(\tilde{d},i,u) \ge (Uv)_i = (Uu)_i - k.$$

By definition of U, $(Uv)_i \ge l(\tilde{d},i,v)$. Hence,

$$(Uu)_i = l(\tilde{d},i,u) \ge (Uv)_i \ge l(\tilde{d},i,v) ,$$

whereby it follows that

$$(Uu)_i - (Uv)_i \le l(\tilde{d},i,u) - l(\tilde{d},i,v)$$
$$\le \rho | u - v | ,$$

since L is contracting. This is valid for all i, hence

$$| Uu - Uv | \le \rho | u - v | . \quad \text{q.e.d.}$$

With this, the conditions of the BANACH Fixed Point Theorem are satisfied. It follows that a fixed point w_δ exists for each decision rule δ. This satisfies the fixed point equation

$$w_\delta = L(\delta, w_\delta) \ . \tag{48.3}$$

In the same manner, there exists exactly one fixed point v^* as the solution to the fixed point equation

$$v^* = Uv^* \ . \tag{48.4}$$

Value iteration is one of the many possibilities for determining v^*. The use of a particular method depends on the extent of the computation.

Let us examine the convergence behavior of value iteration. If one has already performed n iterations, the norm of the residual is reduced after R additional iterations by a factor of ρ^R since

$$\mid v_{n+R} - v^* \mid \ \leq \rho^R \mid v_n - v^* \mid \ . \tag{48.5}$$

How many iterations R are needed to improve the current precision of this approximation v_n by a decimal point? That is,

$$\frac{\mid v_{n+R} - v^* \mid}{\mid v_n - v^* \mid} = \frac{1}{10} \ .$$

From (48.5) one obtains

$$\rho^R = \frac{1}{10}$$

$$\boxed{R = \frac{-1}{\log \rho}} \ .$$

R is a constant. In that case we say that the method is linearly convergent. Looking at it from a numerical standpoint, such methods are tedious. This is true especially if the discount factor ρ lies close to 1. Consider, for example, the following values for a yearly interest of 10%:

Period	ρ	R
Year	0.91	24
Month	0.99	277
Week	0.998	1198

Acceleration of convergence in the discounted case with infinite horizon

A number of methods can be used to accelerate convergence.

1)   Single–step iteration:

Assume that we are computing $v_{n+1}(i)$. For $k < i$, we can use the "better" value $v_{n+1}(k)$ for the required summation instead of $v_n(k)$. This leads to the so–called single–step iteration. The recursion equation is

$$v_{n+1}(i) = \max_{d} \left\{ a_i^d + \rho \sum_{j=1}^{i-1} p_{ij}^d v_{n+1}(j) + \rho \sum_{j=i}^{N} p_{ij}^d v_n(j) \right\} .$$

2)   Further improvements can be brought about by the following variation:

By maximizing with respect to d, the right–hand side of the equation is evaluated for different values of d. If an evaluation leads to an improvement, one uses the improvement $w_{n+1}(i)$ for the next iteration instead of $v_n(i)$.

$$v_{n+1}(i) = \max_{d} \left\{ a_i^d + \rho \sum_{j=1}^{i-1} p_{ij}^d v_{n+1}(j) + \rho p_{ii}^d \max\{v_n(i) \,|\, w_{n+1}(i)\} + \right.$$

$$\left. + \rho \sum_{j=i+1}^{N} p_{ij}^d v_n(j) \right\}.$$

3) Divided Form:

Derivation: We iterate only in the ith–component.

$$w_{n,1}(i) := a_i^d + \rho \sum_{j \neq i} p_{ij}^d w_n(j) + \rho p_{ii}^d w_n(i)$$

$$w_{n,2}(i) := a_i^d + \rho \sum_{j \neq i} p_{ij}^d w_n(j) + \rho p_{ii}^d w_{n,1}(i)$$

$$\vdots$$

$$w_{n,k}(i) := [a_i^d + \rho \sum_{j \neq i} p_{ij}^d w_n(j)] \sum_{r=0}^{k-1} (\rho p_{ii}^d)^r + (\rho p_{ii}^d)^k w_n(i)$$

Since $\rho p_{ii}^d < 1$

$$\lim_{k \to \infty} w_{n,k}(i) = \frac{1}{1 - \rho p_{ii}^d} [a_i^d + \rho \sum_{j \neq i} p_{ij}^d w_n(j)].$$

This leads to the iteration procedure:

$$w_n := (I - \rho P_{\delta,D})^{-1} [a_\delta + \rho(P_{\delta,L} + P_{\delta,U}) w_{n-1}] , \tag{48.6}$$

whereby I is the unit matrix and the matrix P_δ is broken down into a lower triangular matrix $P_{\delta,L}$, an upper triangular matrix $P_{\delta,U}$ and a diagonal matrix $P_{\delta,D}$, i.e.,

$$P_\delta = P_{\delta,L} + P_{\delta,D} + P_{\delta,U} .$$

Iteration (48.6) converges also to a fixed point w_δ which can be shown easily by substitution. Hence, value iteration converges also in divided form

$$v_n(i) = \max_d \left\{ \frac{1}{1 - \rho p_{ii}^d} [a_i^d + \rho \sum_{j \neq i} p_{ij}^d v_{n-1}(j)] \right\} \tag{48.7}$$

to a solution v^* of the optimization problem. Value iteration in divided form can also be performed in the above variations (1) and (2) which lead to an additional acceleration of convergence.

Value iteration in the undiscounted case, infinite planning horizon

With a fixed policy δ, the iteration $w_n = L(\delta, w_{n-1})$ generates a non–convergent sequence starting with $w_0 \equiv 0$,

$$w_n = a_\delta + P_\delta a_\delta + \ldots + P_\delta^{n-1} a_\delta .$$

The change $\Delta n = w_n - w_{n-1}$ tends, however, towards a constant vector. It is

$$\Delta_n = P_\delta^{n-1} a_\delta \tag{48.8}$$

and with stochastic matrices, the limit Π_δ exists for each δ

$$\lim_{n \to \infty} P_\delta^n = \Pi_\delta ,$$

such that the increase Δ_n also has a limit

$$\lim_{n \to \infty} \Delta_n = \lim_{n \to \infty} P_\delta^{n-1} a_\delta . \tag{48.9}$$

Δ is the stationary periodic return which is also called the average return. The numerical problem is that of determining the decision rule δ^* which gives the highest average return Δ^* ($\Delta^* \geq \Delta_\delta$ for all δ, element by element)

$$\Delta^* = \max_\delta \Delta_\delta.$$

This is done by the

Value iteration in the undiscounted case

Step 1:   Start with $v_0 \equiv 0$, $r > 0$.

Step 2:   Compute $v := Uv_0$; compute $\Delta_{old} := v - v_0$.

Step 3:   Compute Uv; compute $\Delta_{new} := Uv - v$.

Step 4:   Is $| \, \| \Delta_{old} \| - \| \Delta_{new} \| \, | > r$?

           Yes: Set $\Delta_{old} := \Delta_{new}$; $v := Uv$

                  and go to 3.

           No: Go to 5.

Step 5:   Stop.

Basically, in the value iteration with infinite planning horizon, one obtains only an approximation of an optimal value v^*. Hence, one can also never be sure that the optimal policy δ^* has been determined. It is possible that one can still achieve a better policy continuing the iteration procedure.

However, there are methods which can recognize and select sub–optimal decision rules partly at the start and partly during the iteration (compare McQUEEN (1967), BARTMANN (1976)). If a single policy still remains then one can be sure that this is also optimal.

§49   POLICY ITERATION

Let us return to the undiscounted case $\rho < 1$ with an infinite horizon. The method of policy iteration proceeds according to the following manner:

> Choose a decision rule δ_1;
>
> compute w_{δ_1};
>
> search for a decision rule δ_2, which improves w_{δ_1};
>
> compute w_{δ_2};
>
> search for a decision rule δ_3, which improves on w_{δ_2}; etc.

In this way a sequence of fixed points w_{δ_i} is computed which is monotonically increasing. Since only a finite number of decision rules exist in the discrete case, this chain stops at a maximum fixed point w_{δ_m} after a finite number of steps.

$$w_{\delta_1} < w_{\delta_2} < \ldots < w_{\delta_m} = \max_{\delta} w_{\delta} .$$

While value iteration is advantageous if a starting vector can be given as a good approximation for v^* (the BANACH Fixed Point Theorem shows that one can begin with only a starting vector $v_o \in \mathbb{R}^N$), policy iteration is recommended if one determines a good approximation for δ^* as a starting decision rule.

Policy iteration in the discounted case
Step 1.   Start with δ.

Step 2.   Compute w_{δ} as a solution to the system of equations

$$w_{\delta} = L(\delta, w_{\delta}) .$$

Step 3.   Test for optimality of δ:

a: Compute Uw_δ. Let the maximizing decision rule be δ'.

b: Is $\delta \neq \delta'$?

- Yes: Let $\delta := \delta'$ and go to 2;

No : Let $\delta^* := \delta$; $v^* := w_\delta$ and go to 4.

Step 4.   Stop.

The termination criterion gives the optimal decision rule δ^* (with value iteration, this is not guaranteed!). It still remains to be proven, however, that the optimal value v^* of the dynamic program can be actually determined in this manner.

Lemma 49.1:   Among the optimal strategies of a Markovian decision problem of the above type with a finite set of decisions is a stationary decision rule δ^*.

Proof:

With an infinite planning horizon, an optimal strategy of the above problem is an infinite sequence $\ldots \delta_n \delta_{n+1} \ldots$ of decision rules (exactly one for each). Either a stationary strategy $\ldots \delta \delta \ldots$ is already optimal or there exists a stationary improvement $\hat{\delta}$ because of the monotonicity of L

$$\ldots \delta \delta \hat{\delta} \hat{\delta} \ldots$$

Since the set of decision rules is finite, there exists a stationary strategy for which there is no stationary improvement.   *q.e.d.*

Undiscounted case.

The decision rule is a bit difficult in this case. We limit ourselves therefore to the so—called complete ergodic case. It implies that the state probabilities

$\pi_{n,\delta}(i)$:   Probability that the system is in state i after period n by using policy δ.

are independent of the initial state for $n \to \infty$.

Let $\pi_{n,\delta} = (\pi_n(1), \pi_n(2),...,\pi_n(N))_\delta$ be the distribution of the state probabilities of the system after n periods beginning with the initial distribution $\pi_{0,\delta}$. Then

$$\pi_{n,\delta} = \pi_{0,\delta} P_\delta^n . \tag{49.1}$$

In the complete ergodic case, the limiting value is

$$\pi_\delta = \lim_{n \to \infty} \pi_{n,\delta}$$

$$= \lim_{n \to \infty} \pi_{0,\delta} P_\delta^n \tag{49.2}$$

independent of $\pi_{0,\delta}$. Since $\lim_{n \to \infty} P_\delta^n = \Pi_\delta$, then for (49.2)

$$\pi_{0,\delta} \Pi_\delta = \pi_\delta . \tag{49.3}$$

Since this relationship must also be valid for the improper initial distributions $(1,0,...,0)$, $(0,1,...,0,0)$, ... $(0,0,...,0,1)$, it follows that the matrix Π_δ has identical rows. Then the average return Δ becomes a vector with identical components (48.9)

$$\Delta_\delta = \Pi_\delta a_\delta = \bar{a}_\delta e , \tag{49.4}$$

$e^T = (1,...,1)$. The average return is then independent of the initial state. It is a scalar of size $\bar{a}_\delta$.

$\bar{a}_\delta$: Average return (stationary periodic return) of a fixed undiscounted ergodic state process;

The total return $v_n(i)$ asymptotically satisfies the relationship

$$v_n(i) = n\bar{a} + V(i), \quad n \text{ very large}. \tag{49.5}$$

Let

V_n: Difference between the expected total return of the n periods and the nth average return $n\bar{a}$

V: $\lim\limits_{n \to \infty} V_n$

The expected return in a single period starting in state i is a_i and for any period n

$$a_i + \sum_j p_{ij} a_j + \dots + \sum_j p_{ij}^{(n-1)} a_j .$$

We write the difference in the form

$$
\begin{aligned}
V_n(i) &= a_i + \sum_j p_{ij} a_j + \dots + \sum_j p_{ij}^{(n-1)} a_j - (n-1)\bar{a} - \bar{a} \\
&= a_i + \sum_k p_{ik} \{ a_k + \sum_j p_{kj} a_j + \dots + \sum_j p_{kj}^{(n-2)} a_j \} - (n-1)\bar{a} - \bar{a} \\
&= a_i + \sum_k p_{ik} v_{n-1}(k) - (n-1)\bar{a} - \bar{a} \\
&= a_i + \sum_k p_{ik} \underbrace{[v_{n-1}(k) - (n-1)\bar{a}]}_{= V_{n-1}(k)} - \bar{a}
\end{aligned}
$$

whereby the following recursion is finally derived

$$V_n(i) + \bar{a} = a_i + \sum_j p_{ij} V_{n-1}(j) \ . \tag{49.6}$$

For $n \to \infty$, this becomes

$$\boxed{V(i) + \bar{a} = a_i + \sum_j p_{ij} V(j) \ .} \tag{49.7}$$

Hence, the resulting value function V measures the total deviation of the total returns from the accumulated average returns. The values $V(i)$ in (49.7) are determined for all but one common factor. For normalization, we set any component to zero, e.g., $V(N)=0$. The other values $V(j)$, $j \neq i$ and $\bar{a}$ are then computed by solving the system of equations (49.7).

How is the best decision rule determined?
It must maximize the stationary single period increase.

For a planning horizon n, it is required to

$$\max_d \left\{ a_i^d + \sum_j p_{ij}^d v_{n-1}(j) \right\} \ .$$

It can also be written as

$$\max_d \left\{ a_i^d + \sum_j p_{ij}^d [n\bar{a} + V_{n-1}(j)] \right\} \ . \tag{49.8}$$

The maximizing d remains the same even if one subtracts the term $n\bar{a}$ (it is independent of d). The test size is then

$$\max_d \left\{ a_i^d + \sum p_{ij}^d V_{n-1}(j) \right\}$$

and for an infinite planning horizon

$$\max_d \left\{ a_i^d + \sum p_{ij}^d V(j) \right\}. \tag{49.9}$$

Policy iteration with $\rho = 1$, complete ergodic case

Step 1:   Start with decision rule δ.

Step 2:   Set $V(N) = 0$;
                compute V und $\bar{a}$ as solution to the system of equations
$$V + \bar{a}e = a_\delta + P_\delta V.$$

Step 3:   Test for optimality of δ:
                a: Compute $\max_\delta \{a_\delta + P_\delta V\}$. Let the maximizing decision rule be δ'.

                b: Is $\delta \neq \delta'$?
                        Yes: Set $\delta := \delta'$ and go to 2;
                        No: Set $\delta^* := \delta; \bar{a}^* := \bar{a}$ and go to 4.

Step 4:   Stop.

§50   BISECTION METHOD AND DYNAMIC PROGRAMMING

The policy iteration has the disadvantage of poorly estimating *a priori* the extent of the computation. The disadvantage of value iteration lies in its very slow convergence if the discount factor is close to 1. Let us, therefore, look at a third method: the Bisection Method in connection with dynamic programming (BARTMANN (1979)).

The bisection method is applied to determine a zero element of a real–valued function. The interval containing the zero element is continually shortened by halving until its length falls into a specified stopping limit. To shorten an interval into one–tenth of its size, aproximately 3.32 bisections are needed. If the bisection method can be applied successfully to a Markovian decision process with an infinite planning horizon and discount factor $\rho < 1$, the computational effort will be independent of ρ and shorter than in value iteration as long as $\rho > 0.5$.

The problem of computing the fixed point v^* is now represented as a problem in $\mathbb{R}^N$. An interval

$$v^- \leq v^* \leq v^+$$

is easy to determine; however, the bisection method does not hold since $\mathbb{R}^N$ is only half–ordered. This means that after the bisection step

$$v_B := \frac{v^+ + v^-}{2}$$

not only $v^* \epsilon [v^-;v_B]$ (situation 1)

or (exclusive or) $v^* \epsilon [v_B;v^+]$ is possible (situation 2) but situation 3 can also occur.

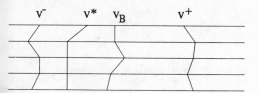

Situation 1: v^* lies in
the left sub–interval

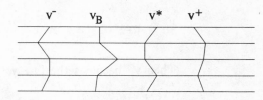

Situation 2: v^* lies in
the right sub–interval

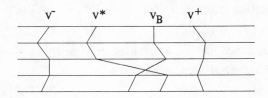

Situation 3: v^* lies neither wholly in the left
nor wholly in the right sub–interval

Explanation of the above diagram:

Each horizontal line means a real axis. Each component $v(i)$ of v is placed on a specific axis. The combination of these individual values results in the jagged line as a representation of the vector v.

To be able to apply the bisection method it must be modified. The whole method consists of 5 parts.

Part 1:   Computation of a suitable starting interval which contains v^*
Part 2:   Bisection step.
Part 3:   Test which of the three situations arise.

Part 4:   If Situation 3 occurs: Perform some maximization steps $v_n := Uv_{n-1}$,
               until monotony is achieved, i.e.,

$$v_n \geq v_{n-1} \text{ or } v_n \leq v_{n-1}.$$

Then v_n becomes the interval partitioning vector, since it follows that
either

a) $v_n > v_{n-1} \Rightarrow v^* > v_n \Rightarrow v^* \in [v_n; v^+]$ or

b) $v_n < v_{n-1} \Rightarrow v^* < v_n \Rightarrow v^* \in [v^-; v_n]$.

Part 5:   Termination criterion.

These parts, when combined, can be written as an algorithm. We formulate it for the
standard situation "Maximization problem, all $a_{ij} \leq 0$ ".

Bisection Method and Dynamic Programming
Step 1 :   Start with $v_0 \equiv 0$, ϵ_{abs}, $\epsilon_{rel} > 0$.
Step 2 :   Compute $v := Uv_0$. Let the maximizing decision rule be δ.
Step 3 :   Set $v^+ := v$ (upper limit of v^*, since $Uv < v$).
Step 4 :   Compute w_δ for δ from Step 2 by solving the system of equations
               $$w_\delta = L(\delta, w_\delta).$$
Step 5 :   Termination criterion:
               a)   Check, if δ is already optimal:
                      compute $\hat{v} := Uw_\delta$;

                      if δ is again the maximizer,
                      set $v^* := \hat{v}$; $\hat{\delta}^* := \delta$ and go to 14.
               b)   if $|\hat{v} - w_\delta| / |\hat{v}| < \epsilon_{rel}$,
                      set $v^* := \hat{v}$; $\hat{\delta}^* := \delta$ and go to 14.

Step 6 :   Set $v^- := \hat{v}$ (lower limit of v^*).

Step 7 :   Bisection step: $v_B := (v^+ + v^-)/2$

Step 8 :   Termination criterion: $|v^+ - v^-| < \epsilon_{abs}$?

            Yes: compute Uv, to obtain δ^*,

                  set $v^* := Uv$; $\delta^* := \delta$ (maximizer of Uv)

                  and go to 14;

            No :   go to 9.

Step 9 :   Test, which situation arises:

            Compute Tv (corresponds to Uv with possible computational reduction)

                 a) if $Tv \geq v$, set $v^- := v$ and go to 7;

                 b) if $Tv \leq v$, set $v^+ := v$ and go to 7;

                 c) if $Tv \equiv v$, set $v^* := Tv$ and go to 14;

                 d) otherwise, go to 10.

Step 10:   Situation 3 arises:

            Compute $\hat{v} := Uv$. Let the maximizer be δ.

Step 11:   Termination criterion:

                 If $|\hat{v} - v|/|\hat{v}| < \epsilon_{rel}$, set $v^* := \hat{v}$; $\delta^* := \delta$

            and go to 14.

Step 12:   Test, whether monotony arises:

                 a) if $\hat{v} \geq v$, set $v^- := \hat{v}$; $\delta^* := \delta$

            and go to 7;

                 b) if $\hat{v} \leq v$, set $v^+ := \hat{v}$; $\delta^* := \delta$

            and go to 7;

                 c) otherwise, go to 13.

Step 13:   Set $v := \hat{v}$ and go to 10.

Step 14:   Stop.

Explanation of iteration Tv in Step 9: To test the three situations mentioned above, one can perform a full maximization step Uv and compare v with Uv. In order to satisfy any one of the conditions: (a) $v^* < v$;   (b) $v^* > v$;   or   (c) $v^* \gtrless v$, a full maximization step is unnecessary in all cases. For example, to test for $v^* > v$, it is sufficient to determine a decision rule δ which results into an improvement $L(\delta, v) > v$.

(Note: This is a maximization problem, hence " > " is used). One need not look for the best rule. Similarly, to determine the occurrence of Situation 3, it is sufficient to determine two components i, j such that $v_i < (Uv)_i$ and $v_j > (Uv)_j$. The other components need no longer be tested anymore. One can therefore define Tv in the following manner.

Iteration Tv:

First Component:
For the valid decisions in state 1, compute the size $d(d,1,v)$. As soon as $l(d,1,v) > v(1)$ for a given d compute for the other components.

Remaining Components:
a)   Let a d be found for which $l(d,1,v) > v(1)$. One can stop computing for the other components i as soon as one determines a d such that $l(d,1,v) > v(i)$. If no such decision exists in a state $j > 1$, i.e., $\max_d(d,j,v) < v(j)$, then it follows immediately that $Uv \lessgtr v$ (not comparable) and one can stop the test.

b)   Let no d be found for which $l(d,1,v) > v(1)$. Perform the full maximization step in the remaining components. However, if a d is found in a state $i > 1$ such that $l(d,1,v) > v(i)$, then it follows immediately that $Uv \lessgtr v$ and one can stop the test.

The above bisection method in combination with dynamic programming is a basic algorithm which allows numerous variations and fine–tuning. Numerical experience shows that this method performs a lot better than value and decision iteration.

If a very special problem arises, methods tailored to such situations can be very effective. One example is described in the next section.

§51 COMPUTATION OF OPTIMAL (s, S)–POLICIES ACCORDING TO FEDERGRUEN/ZIPKIN

The method of FEDERGRUEN/ZIPKIN for the computation of optimal (s, S) policies is related to the undiscounted standard AHM model with backorders. The distribution of U is in the form $p_0, p_1, p_2, p_3, \ldots$ We consider the model with proportional ordering cost. Let

$\hat{f}(y)$: Expected value of inventory holding and shortage costs of a period with starting inventory y

y : Initial inventory (before a possible order)

x : Initial inventory after an order

x–y : Order quantity

δ : Ordering rule of the (s,S)–Type

$$\delta(y) = \begin{cases} y, & \text{if } s+1 \le y \le S; \\ S, & \text{if } y \le s; \end{cases}$$

δ^* : Optimal ordering rule

c_δ : Average cost for an ordering rule δ

c^* : Minimum average cost

π_y^δ : Stationary state probability of inventory y with policy δ

F(x,y) : Single period cost

k : Fixed ordering cost

$$F(x,y) = \begin{cases} \hat{f}(y), & \text{if } x = y; \\ k + \hat{f}(y), & \text{if } x > y; \end{cases}$$

$$F_\delta(y) = F(\delta(y),y) .$$

The Principle of Optimality is expressed using the following notation

$$v(y) + c^* = F_\delta(y) + \sum_{u=0}^{\infty} v[\delta(y) - u] p_u \, , \tag{51.1}$$

for all y < S. By definition

$$v(S) = 0 \tag{51.2}$$

for normalization. If one wants to solve the functional equations (51.1) we must restrict the state space to a finite size, i.e., allow a smallest $y = y_{min}$ such that $y_{min} \le y \le S$. y_{min} serves as an absorbing barrier. Equation (51.1) is changed accordingly. The summation should be performed until $\delta(y) - u = y_{min}$. This restriction of the state space causes an inaccuracy in the model.

The method of FEDERGRUEN/ZIPKIN avoids this. It is not based on the recursive evaluation of the functional equations but monitors the stock movement after inventory is filled up to S.

We define

t(w)   :   Expected time until the next order, if the current stock lies w units above the ordering point, $w = y - S$, w > 0.

$\nu_s(y)$:   Expected costs until the next order, if the current stock is y, y > s.

Both functions t and ν satisfy the equations

$$t(w) = 1 + \sum_{u=0}^{w-1} p_u t(w - u), \qquad w > 0 \, , \tag{51.3}$$

$$\nu_s(y) = \hat{f}(y) + \sum_{u=0}^{y-s-1} p_u \nu_s(y - u), \qquad y > s. \tag{51.4}$$

t is independent of the (s, S) Policy and ν depends only on s. The system of equations (51.3) has a triangular form:

$$t(1) - 1 = p_0 t(1).$$
$$t(2) - 1 = p_0 t(2) + p_1 t(1)$$
$$\vdots \qquad \ddots$$
$$t(w) - 1 = p_0 t(w) + \cdots + p_{w-1} t(1) .$$

Similarly, the system of equations (51.4) has the same form.

Starting with w = 1, t can be computed very rapidly. The same is true for ν_s, starting with y = s + 1. The essential advantage of the method lies in the fact that one can compute the values below with t and ν

$$c_\delta = \frac{\nu_s(S) + k}{t(S - s)} \tag{51.5}$$

$$v_\delta(y) = \begin{cases} \nu_s(y) + k - c_\delta t(y - s), & \text{for} \quad y > 0 \\ k & , \quad \text{for} \quad y \le s \end{cases} \tag{51.6}$$

Equation (51.5) is exactly the cycle cost $\nu_s(S) + k$ per cycle. The validity of (51.5), (51.6) shows that these expressions satisfy the Principle of Optimality when substituted in (51.1), (51.2). Thus a fast method of policy iteration can be constructed with equations (51.3) to (51.6).

Step 1: Initialization.

Set limits $\underline{s}$, $\underline{S}$, $\overline{S}$ for the values s, S.

$\underline{s}$: smallest whole number such that $\hat{f}(\underline{s}) \le \hat{f}(\underline{S}) + k$;

$\underline{S}$: smallest whole number which minimizes $\hat{f}(y)$;

$\overline{S}$: smallest whole number such that $\hat{f}(\overline{s}) \ge \hat{f}(\underline{S}) + k$; (compare § 42)

Let $s_{old} := S_{old} := -1$

Choose a starting policy $\delta = (s, S)$ and let $s_{new} := s$; $S_{new} := S$.

Compute the function t(w), w = 1, 2 ..., $\overline{S} - \underline{s}$ from (51.3).

Step 2:   Computation of the Value function.

If s has changed from the last iteration ($s_{old} \neq s_{new}$);

Compute $\nu_s(y)$, $y = s + 1$, ... U from equation (51.4).

Compute c_δ and $v_\delta(y)$, $y = \underline{s}$, ..., $\overline{S}$ from equation (51.5), (51.6).

Step 3:   Policy Improvement.

a) Save the old policy: $s_{old} := s_{new}$; $S_{old} := S_{new}$.

b) Compute the minimizing S'; $\underline{S} \leq S' \leq \overline{S}$

$$v_\delta(S') = \min_{\underline{S} \leq y \leq \overline{S}} v_\delta(y) \,.$$

$S_{new} := S'$.

c) Search for a better s:

c1)   in increasing order: $s+1$, $s+2$,...,$\overline{s}$;

in case an order is worthwhile at state $s+1$, i.e., if

$$k + v_\delta(S') < v_\delta(s + 1) \,,$$

search further in increasing order until a y is found for which it is no longer profitable to order. Let this case be at inventory level η. It must therefore follow that

$$k + v_\delta(S') < v_\delta(y) \quad \text{for all } y, \quad s < y \leq \eta - 1 \,.$$

Set $s_{new} := \eta - 1$.

Go to Step 4.

c2)   in decreasing order: s–1, s–2, ..., $\underline{s}$;

in case an order is not worthwhile at state s–1, i.e., if

$$\hat{f}(s - 1) < c_{\delta,}$$

search further in decreasing order until a y is found for which it is profitable to order. Let this case be at inventory level ξ. It must therefore follow that

$$\hat{f}(y) < c_\delta \quad \text{for all y,} \quad \xi + 1 \leq y < s.$$

Set $s_{new} := \xi + 1$.

Go to Step 4.

Step 4:   Termination Criterion.

If $\delta_{old} = (s_{old}, S_{old}) \neq \delta_{new} = (s_{new}, S_{new})$, let $\delta_{new} := \delta_{old}$ and go to step 1. If $\delta_{old} = \delta_{new}$ then go to 5. δ_{new} is the optimal policy.

Step 5.   Stop.

All methods discussed in this chapter (exluding value iteration) for inventory problems of realistic sizes can be computed within seconds on a personal computer.

CLOSING REMARKS

Inventory theory is by no means exhausted. A book can also never hope for completeness. It only presents a selection and thus is quite subjective. Our objectivce was to present the most important and typical methods and models in order to stimulate the reader to reflect on his own.

LITERATURE

Bibliography

A bibliography of works done before 1953 may be found in:
WHITIN, T.M.: The Theory of Inventory Management, Princeton 1953.

For the period 1953–55:
GOURARY, M., LEWIS, R., NEELAND, F.: An Inventory Control Bibliography.
Naval Research Logistic Quarterly, 3(1955), 295–304.

Abstracts of works in English for the period 1953–65 are found in: EILON, S.,
LAMPKIN, W., Inventory Control Abstracts.
Edinburgh–London 1968.

A bibliography of stochastic inventory models up to 1967 can be found in:
HOCHSTÄDTER, D.: Stochastische Lagerhaltungsmodelle. Lecture Notes in
Operations Research and Mathematical Economics, Berlin 1968.

RICHARDS, F.R., MARSHALL, K.,T.: The OR/MS Index 1952–1976.
A cumulative Index of Management Science 1–22, Operations Research 1–24, and
Interfaces 1–6. The Institute for Management and Operations Research Society of
America. Providence und Baltimore 1978.

An on–going collection of abstracts are offered by:
BRADLEY, H., (Hrsg.): International Abstracts in Operations Research. Amsterdam,
und

ROSENTHAL, A. (Hrsg.): Operations Research/Management Science. International
Literature Digest. Whippany, New Jersey.

The latest bibliography comes from the International Society for Inventory Research (ISIR):
ATTILA CHIKAN (Hrsg.): Bibliography of Inventory Literature.
ISIR Sekretariat Veres Palne u. 36, Budapest, Hungary, H–1053, 1988.

Monographs

ARROW,K., KARLIN, J., SCARF, H.(ed): Studies in the Mathematical Theory of Inventory and Production. Stanford 1958.

BECKMANN, M.J.: Dynamic Programming of Economic Decisions. Heidelberg 1968.

BEMELMANS, R.: The Capacity Aspect of Inventories. Lecture Notes in Economics and Mathematical Systems, Berlin 1968.

BROWN, R.G.: Decision Rules for Inventory Management. New York 1967.

BUCHAN, J., KOENIGSBERG, E.: Scientific Inventory Management.
Englewood Cliffs N.J. 1963.

BUFFA, E.: Production–Inventory systems: Planning and Control.
Homewood 1968.

HADLEY, G., WHITIN, T.M.: Analysis of Inventory Systems. Englewood Cliffs N.J. 1963.

HANSSMANN, F.: Operations Research in Production and Inventory Control. New York 1962.

HOCHSTÄDTER, D.: Stochastische Lagerhaltungsmodelle. Lecture Notes in Operations Research and Mathematical Economics, Berlin 1969.

HOLT, C.,C., MODIGLIANI, F., MUTH, J.F., SIMON, H.A.: Planning Production, Inventories and Work Force. Englewood Cliffs, N.J. 1960.

MAGEE, J.F., BOODMANN, D.: Production Planning and Inventory Control (2nd ed), New York 1967.

SCHNEEWEISS, CH.: Inventory Production Theory. A Linear Policy Approach. Lecture Notes in Economics and Mathematical Systems. Berlin 1978.

STARR, M., MILLER, D.: Inventory Control: Theory and Practice. Englewood Cliffs, N.J. 1962.

TERSINE, R.J.: Principles of Inventory and Materials Management. New York (2nd ed.) 1982.

TIJMS, H.C.: Analysis of (s,S) Inventory Models. Amsterdam 1972.

Books and Articles

ALSCHER, J., KÜHN, M., SCHNEEWEISS, CH.: On the validity of reorder point inventory models for regular and sporadic demand. Engineering Costs and Production Economics 10 (1986), pp. 43 ff.

ARROW, K.J., HARRIS, T., MARSCHAK, J.: Optimal Inventory Policy. Econometrica 19 (1951) 3, pp. 250 ff.

AXSAETER, S., SCHNEEWEISS, Ch., SILVER, E. (Ed.): Multi–Stage Production Planning and Inventory Control. Lecture Notes in Economics and Mathematical Systems, Berlin 1986.

BARTMANN, D.: A Method of Bisection for Discounted Markov Decision Problems. Zeitschrift für Operations Research 23 (1979), pp. 275 ff.

BARTMANN, D.: Optimierung Markovscher Entscheidungsprozesse. Dissertation, TU München 1976.

BARTMANN, D.: Optimierung eines Zweiproduktlagers. In: Dathe, H. (Hrsg.), Proceedings in Operations Research 6. Würzburg–Wien 1976.

BECKMANN, M.J.: On the Theory of Stochastic Control Processes. Bull. Soc. Royale Sciences Liege 33 (1964), pp.. 520–529.

BECKMANN, M.J.: Dynamic Programming and Inventory Control. Operations Research Quarterly 15 (1964) 4, pp. 389–400.

BECKMANN, M.J.: Production Smoothing and Inventory Control. Operations Research 9 (1961), pp. 456–467.

BECKMANN, M.J., HOCHSTÄDTER, D.: Berechnung optimaler Entscheidungs– regeln für die Lagerhaltung. Jahrbücher für Nationalökonomie und Statistik, 182 (1968), pp. 106.123.

BELLMAN, R.: Dynamic Programming. Princeton N.J. 1957.

BERNOULLI, D.: Versuch einer neuen Theorie der Wertbestimmung von Glücksfällen. In: Pringsheim, A. (Hrsg.), Die Grundlagen der modernen Wertlehre, Leipzig 1896.

BLACKBURN, J.D., MILLEN, R.A.: A heuristic lot–sizing performance in a rolling–schedule environment. Decision Sciences 11 (1980), pp.691 ff.

BOX, G.E.P., JENKINS, G.M.: Time Series Analysis: Forecasting and Control, San Francisco 1976.

BUEHLER, G.: Sicherheitsäquivalente und Informationsbedarf bei stochastischen dynamischen Produktions–Lagerhaltungs–Modellen. Frankfurt 1979.

CHANAL, S.: A note on dynamic lot sizing in a rolling–horizon environment. Decision Sciences 13 (1982) 1, pp. 113 ff.

COLLATZ, L.: Funktionalanalysis und numerische Mathematik. Heidelberg 1968.

DE GROOT, M.H.: Optimal Statistical Decisions. New York 1970.

DE MATTEIS, J.J., MENDOZA, A.G.: An Economic Lot–sizing Technique. IBM Systems Journal 7 (1968), S. 30 ff.

D´EPENOUX, F.: Sur un probleme de production et de stockage dans l´aleatoire. Revue Francaise de recherche operationnnelle (1960), S. 3–15.

D´EPENOUX, F.: A Probabilistic Production and Inventory Problem. Management Science, Vol 10 (1963), pp. 98–108.

FEDERGRUEN, A., und ZIPKIN, P.: An Efficient Algorithm for Computing Optimal (s,S) Policies. Operations Research 32 (1984) 6, pp. 1268 ff.

GRUBBSTRÖM, R.W.: Dynamical Aspects of Production Inventory Systems. Third International Symposium on Inventories, Budapest 1984.

IGLEHART, D.L.: Optimality of (s, S) Policies in the Infinite Horizon Dynamic Inventory Problem. Management Science 9 (1963) 2, pp. 259–267.

HOCHSTÄDTER, D.: Neuere Entwicklung der stochastischen Lagerhaltungstheorie. In: Beckmann, M.J.(Hrsg.), Unternehmensforschung heute. Lecture Notes for Operations Research and Mathematical Systems, Berlin 1971.

HOWARD, R.A.: Dynamic Probabilistic Systems, Vol.I and II. New York, 1971.

KNOLMAYER, G.: Ein Vergleich von 30 "praxisnahen" Lagerhaltungsheuristiken. In: Ohse, D. u.a. (Hrsg), Operations–Research–Proceedings 1984, Berlin 1985, pp. 223 ff.

McQUEEN, J.B.: A Test for Suboptimal Actions in Markovian Decision Problems. Operations Research 15 (1987) 3, pp. 559 ff.

MEYER, M., HANSEN, K.: Mathematische Planungsverfahren II. Eine einführende und anwendungsorientierte Darstellung von Lagerhaltungs– und Warteschlangenmodellen. Essen 1975.

OHSE, D.: Näherungsverfahren zur Bestimmung der wirtschaftlichen Bestellmenge bei schwankendem Bedarf. Elektronische Datenverarbeitung 12 (1970), pp. 83 ff.

PREKOPA, A. (Hrsg.): Inventory Control and Water Storage. Amsterdam 1973.

REINFELD, N.V. (Ed.): Handbook of Production and Inventory Control. New Jersey 1987.

RENYI, A.: Briefe über die Wahrscheinlichkeit. Basel 1969.

SASIENI, M., JASPAN, A., FRIEDMAN, L.: Methoden und Probleme der Unternehmensforschung. Würzburg 1962.

SCARF, H.E., GILFORD, D.M., SHELLY, M.W. (Ed.): Multistage Inventory Models and Techniques. Stanford 1963.

SCARF, H.: The Optimality of (S,s) Policies in Dynamic Inventory Problems. In: Arrow, K., Karlin, Suppes, P. (Hrsg.), Mathematical Methods in the Social Sciences, 1959. Proceedings of the first Stanford Symposium. Stanford 1960, pp. 196–202.

SCHÄL, M.: On the Optimality of (s,S)–Policies in Dynamic Inventory Models with Finite Horizon. Siam Journal of Applied Mathematics 30 (1976), pp. 528 ff.

SCHNEEWEISS, CH.: Optimal Production Smoothing and Safety Inventory. Management Science 20 (1974), pp. 1122–1130.

SILVER, E.A., MEAL, H.C.: A Heuristic for Selecting Lot Size Quantities for the Case of a deterministic time–varying Demand Rate and discrete Opportunities for Replenishment. Production and Inventory Management 14 (1973) 2, S. 64 ff.

252

SILVER, E.A., MILTENBURG, J.: Two Modifications for the Silver–Meal Lot Sizing Heuristic. Infor 22 (1984) 1, S. 56 ff.

SIMON, H.: Dynamic Programming under Uncertainty with a Quadratic Criterion Function. Econometrica 24 (1956), S. 74–81.

VEINOTT, A.: On the Optimality of (s,S) Inventory Policies: New Conditions and a New Proof. Siam Journal of Applied Mahtematics 14 (1966) S. 1067–1083.

VEINOTT, A., WAGNER, H.: Computing Optimal (s,S) Inventory Policies. Management Science 11 (1965), S. 525–552.

WAGNER, H.M., WHITIN, T.M.: Dynamic Version of the Economic Lot Size Model. Management Science 5 (1958), S. 89 ff.

Printing: Weihert-Druck GmbH, Darmstadt
Binding: Buchbinderei Schäffer, Grünstadt

Lecture Notes in Economics and Mathematical Systems

For information about Vols. 1–210
please contact your bookseller or Springer-Verlag

Vol. 359: E. de Jong, Exchange Rate Determination and Optimal Economic Policy Under Various Exchange Rate Regimes. VII, 270 pages. 1991.

Vol. 360: P. Stalder, Regime Translations, Spillovers and Buffer Stocks. VI, 193 pages . 1991.

Vol. 361: C. F. Daganzo, Logistics Systems Analysis. X, 321 pages. 1991.

Vol. 362: F. Gehrels, Essays In Macroeconomics of an Open Economy. VII, 183 pages. 1991.

Vol. 363: C. Puppe, Distorted Probabilities and Choice under Risk. VIII, 100 pages . 1991

Vol. 364: B. Horvath, Are Policy Variables Exogenous? XII, 162 pages. 1991.

Vol. 365: G. A. Heuer, U. Leopold-Wildburger. Balanced Silverman Games on General Discrete Sets. V, 140 pages. 1991.

Vol. 366: J. Gruber (Ed.), Econometric Decision Models. Proceedings, 1989. VIII, 636 pages. 1991.

Vol. 367: M. Grauer, D. B. Pressmar (Eds.), Parallel Computing and Mathematical Optimization. Proceedings. V, 208 pages. 1991.

Vol. 368: M. Fedrizzi, J. Kacprzyk, M. Roubens (Eds.), Interactive Fuzzy Optimization. VII, 216 pages. 1991.

Vol. 369: R. Koblo, The Visible Hand. VIII, 131 pages.1991.

Vol. 370: M. J. Beckmann, M. N. Gopalan, R. Subramanian (Eds.), Stochastic Processes and their Applications. Proceedings, 1990. XLI, 292 pages. 1991.

Vol. 371: A. Schmutzler, Flexibility and Adjustment to Information in Sequential Decision Problems. VIII, 198 pages. 1991.

Vol. 372: J. Esteban, The Social Viability of Money. X, 202 pages. 1991.

Vol. 373: A. Billot, Economic Theory of Fuzzy Equilibria. XIII, 164 pages. 1992.

Vol. 374: G. Pflug, U. Dieter (Eds.), Simulation and Optimization. Proceedings, 1990. X, 162 pages. 1992.

Vol. 375: S.-J. Chen, Ch.-L. Hwang, Fuzzy Multiple Attribute Decision Making. XII, 536 pages. 1992.

Vol. 376: K.-H. Jöckel, G. Rothe, W. Sendler (Eds.), Bootstrapping and Related Techniques. Proceedings, 1990. VIII, 247 pages. 1992.

Vol. 377: A. Villar, Operator Theorems with Applications to Distributive Problems and Equilibrium Models. XVI, 160 pages. 1992.

Vol. 378: W. Krabs, J. Zowe (Eds.), Modern Methods of Optimization. Proceedings, 1990. VIII, 348 pages. 1992.

Vol. 379: K. Marti (Ed.), Stochastic Optimization. Proceedings, 1990. VII, 182 pages. 1992.

Vol. 380: J. Odelstad, Invariance and Structural Dependence. XII, 245 pages. 1992.

Vol. 381: C. Giannini, Topics in Structural VAR Econometrics. XI, 131 pages. 1992.

Vol. 382: W. Oettli, D. Pallaschke (Eds.), Advances in Optimization. Proceedings, 1991. X, 527 pages. 1992.

Vol. 383: J. Vartiainen, Capital Accumulation in a Corporatist Economy. VII, 177 pages. 1992.

Vol. 384: A. Martina, Lectures on the Economic Theory of Taxation. XII, 313 pages. 1992.

Vol. 385: J. Gardeazabal, M. Regúlez, The Monetary Model of Exchange Rates and Cointegration. X, 194 pages. 1992.

Vol. 386: M. Desrochers, J.-M. Rousseau (Eds.), Computer-Aided Transit Scheduling. Proceedings, 1990. XIII, 432 pages. 1992.

Vol. 387: W. Gaertner, M. Klemisch-Ahlert, Social Choice and Bargaining Perspectives on Distributive Justice. VIII, 131 pages. 1992.

Vol. 388: D. Bartmann, M. J. Beckmann, Inventory Control. XV, 252 pages. 1992.